Real Science-4-Kids

SCIENCE

Book 4

Rebecca Woodbury, Ph.D., M.Ed.

RS4K

RS4K Real Science-4-Kids

Illustrations: Janet Moneymaker

The Building Blocks Series

Science Book 4 Student Textbook (hardcover)
ISBN 978-1-941181-88-1

Published by Gravitas Publications Inc.
Real Science-4-Kids®
www.gravitaspublications.com

Table of Contents

Physics

Geology

Astronomy

CHAPTER 1

Discovering with SCIENCE

1.1 Introduction

What is a **discovery**? Where do discoveries come from? Why do we delight in discoveries?

In science, a discovery is understanding or seeing something that we previously didn't understand about how the world works.

Scientific discoveries are exciting not only because we begin to understand how things work but because they open doors to new ways of doing things. For example, discovery of electrons has made cell phones, computers, automobiles, and indoor lighting possible.

Discoveries in science happen every day. Some discoveries are small and hardly noticed, but other discoveries are big and can get the whole world talking. The discovery of a new planet or a small atomic particle can transform lives and change the future.

In this book we will take a look at some of the discoveries scientists have made in **chemistry**, **biology**, **physics**, **geology**, and **astronomy**.

1.2 Discoveries in Chemistry

One of the most important discoveries in chemistry is the discovery of the **atom**. The atom is the basic building block of all matter. Atoms fit together to make everything we can see, touch, taste, smell, and hear. But discovering the atom was no easy task. It took several hundred years and many different ideas and experiments to finally show that atoms exist.

In the chemistry section of this book you will learn about how atoms bond to make **molecules**, and you will also learn about the nature of **chemical reactions**.

By knowing about atoms, we can explain what happens when ice melts and becomes water and how diamonds and rubies form. By understanding the nature of the atom, we can explain why a homemade pie smells wonderful and why getting sprayed by a skunk smells awful!

Because we understand about atoms and how atoms combine to make molecules during chemical reactions, we can create all kinds of new materials, like plastics, nylon, and even light-weight carbon fibers to make fast bicycles!

1.3 Discoveries in Biology

How many animals are there on Earth? How many types of animals live in oceans or in the forest, on mountain tops or in the desert?

One of most exciting discoveries in science is the discovery of different types of animals. Although many animals have been known for centuries, scientists are still discovering new animals.

In the biology section you will learn about different types of animals, from the ocean sponge to spiders, lizards, and humans! You will learn about how scientists try to define life and why cows are in a different group than mountain lions. You will also learn some very important scientific words like **vertebrate** and **omnivore**.

You will also learn about the process of **metamorphosis** in which some animals, like butterflies and frogs, change form during their life cycle.

1.4 Discoveries in Physics

Some of the most interesting discoveries in science come from physics. Did you know that it took thousands of years and many men and women of science to figure out the nature of light? Now that we understand light, we can make flashlights, lasers, and fiber optics.

In the physics section you will learn how scientists discovered the nature of **waves**. You will also learn about different types of waves and what they look like.

You will learn about the nature of sound, how sound travels at different speeds through different materials, and what happens when sound bounces.

You will also learn about light, the speed of light, and what makes light so special. And you will also see how the physics of light is connected to the chemistry of light as you discover what makes fireworks so beautiful!

1.5 Discoveries in Geology

Discovering the way Earth works is a big job! There is so much about Earth to discover because Earth is so BIG!

Early explorers probably didn't think about Earth in the same way we do today. Over many hundreds of years, we have discovered that Earth works like one big system with thousands of little interconnected systems.

One of the most important discoveries about Earth is how the various **cycles** keep Earth both stable and constantly changing. In the geology section you will learn about the **atmospheric cycle, rock cycle**, and **energy cycle** and how without these carefully balanced cycles, Earth would not be a planet that could support life.

You will also learn about **ecosystems** and how Earth's plants, animals, and geological cycles create specialized communities and areas where only certain plants and animals thrive.

And finally, you will learn about Earth's **natural resources** and how using natural resources has given humans the ability to create amazing technology but has also harmed some ecosystems. You will see that some of the most important discoveries in geology will involve figuring out how to keep all the parts of Earth in balance!

1.6 Discoveries in Astronomy

And what about going beyond Earth and exploring outer space? What can we find on other planets or in deep space?

In the astronomy section you will find out how we have collected information about the universe and how we have created technology to put women and men in space.

You will learn about manned space travel and the process required to get a space project launched. You will learn about **rockets**, how they are designed and the type of fuel needed to launch both humans and equipment into space.

You will also learn about how astronomers are looking for life on other planets in our solar system, in our galaxy and beyond. You will find out how astronomers have discovered planets that could be like Earth and how they can search for life.

Finally, you'll explore the future of astronomy and how new discoveries today will help astronomers make future discoveries.

1.7 Summary

- Making new discoveries is an exciting part of doing science.
- Discoveries happen in every field of scientific study
- As science continues to expand, scientists continue to discover new information about the world we live in.

1.8 Vocabulary

Basic Words

rocket (RAH-kuht)

Challenge Words

astronomy (uh-STRAH-nuh-mee)

atmospheric cycle (aat-muh-SFIHR-ihk SIY-kuhl)

biology (biy-AH-luh-jee)

chemical reaction (KEH-muh-kuhl ree-AAK-shuhn)

chemistry (KEH-muh-stree)

discovery (dih-SKUH-vuh-ree)

ecosystem (EE-koh-sih-stuhm)

energy cycle (EH-nuhr-jee SIY-kuhl)

geology (jee-AH-luh-jee)

metamorphosis (meh-tuh-MAWR-phuh-suhs)

natural resource (NAA-chuh-ruhl REE-sawrss)

omnivore (AHM-nih-vawr)

physics (FIH-zihks)

vertebrate (VUHR-tuh-bruht)

CHAPTER 2

Giving, Taking, Sharing ELECTRONS

2.1 Introduction

In this chapter we'll find out more about how **atoms** combine to make **molecules**. Remember that when two or more atoms meet, they sometimes combine to make a molecule. The way they combine depends on the number and type of **electrons**. In this chapter we will explore how atoms give, take, and share electrons.

2.2 Inside the Atom

Why is it that when a **sodium** atom meets a **chlorine** atom they will join to make table salt, but if a sodium atom meets another sodium atom, they don't join and instead stay away from each other?

What makes each kind of atom unique, and why do some atoms like to meet and join, while other atoms stay away from one another?

Table Salt
Sodium and chlorine atoms join

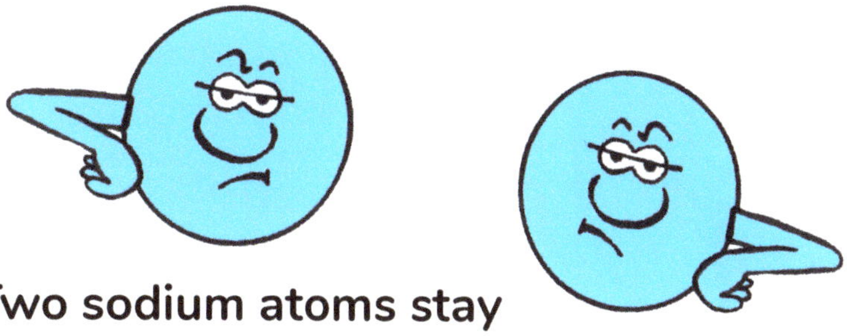

Two sodium atoms stay away from each other.

2.3 What Really Happens When Atoms Meet and Join?

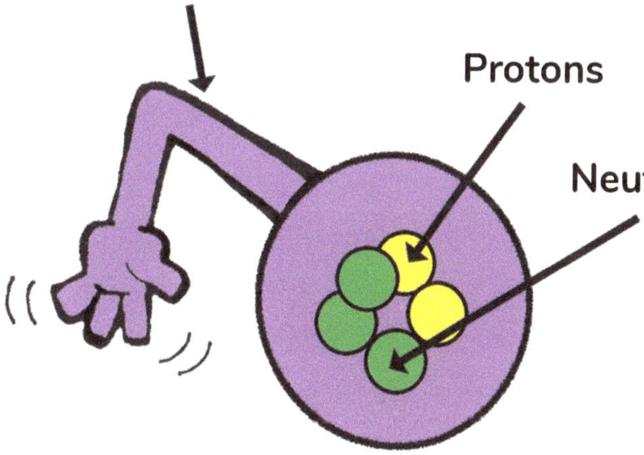

Recall that each atom is made of smaller parts called **protons**, **neutrons**, and **electrons**. The number of protons, neutrons, and electrons an atom has determines the kind of atom it is. Also recall that protons and neutrons are in the center of the atom and electrons occupy the space surrounding the center. Some electrons are available for bonding and some electrons are not available for bonding. One way to illustrate electrons that can form bonds is to use a model that shows the electrons available for bonding as "arms."

This model is useful for demonstrating how atoms hook to each other to make molecules. But it's not useful for showing why some atoms join together and others don't. To explain why some atoms join together and others don't, we need to change our model.

Instead of using an arm to show an electron that is available for bonding, we will show the available electron as a red dot. Red dots will represent electrons in one atom that can hook to electrons in other atoms.

Some atoms have electrons that don't form bonds, and these electrons we'll show as gray dots.

For example, **helium** has two electrons, but neither of these electrons can form bonds. Notice that the two electrons are paired. When electrons form pairs, they don't bond, so in this model they are colored gray.

Lithium has three electrons but only one of those electrons can form a bond. The other two electrons are paired, so they don't form bonds. Notice that in this model, the two paired electrons (in gray) are drawn on the inside of an outer circle, and the red electron is drawn on the outside of the outer circle. This is because the paired gray electrons are like the electrons in helium and have less energy than the single red electron. Putting the gray electrons on the inside of an outer circle is one way to show this.

Beryllium has four electrons. Two of the electrons are paired and, like lithium, have less energy and are on the inside of the outer circle. The other two electrons are single and red. How many bonds can beryllium form? Two! Because it has two red electrons. You may wonder why these two red electrons don't form a pair. It's difficult to explain, but they don't. It's like they are in the same house (on the outside of the outer circle) but they are in different rooms (different places on the outer circle) and so don't form a pair.

Boron has five electrons. Two paired gray electrons inside the outer circle and three red electrons on the outside. How many bonds can boron form? Three! Because it has three red electrons.

Carbon has six electrons. Two paired gray electrons inside the outer circle and four red electrons on the outside. How many bonds can carbon form? Four! Because it has four red electrons.

Nitrogen has seven electrons. Two paired gray electrons inside the outer circle and five electrons on the outside. But two of the electrons on the outside have paired up. Now that they are paired, they can't form bonds. So how many bonds can nitrogen form? Three! Because it has three red electrons.

Oxygen has eight electrons. Two paired gray electrons inside the outer circle and six on the outside. But four of the electrons on the outside have paired up, so there are only two red electrons. How many bonds can oxygen form? Two! Because it has only two red electrons.

Fluorine has nine electrons. Two paired gray electrons inside the outer circle and seven on the outside. But six of the electrons on the outside have paired up, so there are three pairs of electrons, and only one red electron. How many bonds can fluorine form? One! Because it has only one red electron.

Neon has ten electrons. Two paired gray electrons inside the outer circle and eight electrons on the outside. All of electrons for neon are paired, so neon can't form any bonds.

Electrons like to be in pairs, and the way atoms like to hook together is to pair their electrons. Notice that when the electrons within an atom form a pair, those electrons no longer want to form bonds. The same thing happens when an atom bonds to another atom. The electrons form a pair to make the bond, and then they don't want to make more bonds.

2 hydrogen atoms ⟶ join ⟶ to form a bond

To get pairs of electrons, atoms will give away, take, or share electrons.

In a chemical reaction where **electrons are shared**, the bond is called a **covalent bond**. In a covalent bond, the electrons are shared equally. Two hydrogen atoms form a covalent bond when they join to make a hydrogen molecule.

In a chemical reaction where electrons are either given away by an atom or taken from an atom, the bond is called an **ionic bond**. In an ionic bond the **electrons are not shared**. The electrons hang out mostly on the atom that has taken the electrons, and the atom the electrons have been taken from is happy just to be nearby.

2.4 Which Atoms Are Givers, Which Are Takers, and Which Like to Share?

Lithium **Fluorine**

Ionic Bond
(lithium fluoride)

Atoms that end with "ium," such as sodium, lithium, magnesium, calcium, cesium, and beryllium, like to **give their electrons away**. These atoms would rather form an ionic bond by giving away their electrons to another atom than take electrons from another atom to make pairs.

Atoms than end in "ine," such as fluorine, chlorine, bromine, and iodine, like to **take electrons**. These atoms prefer to make all of their electrons into pairs by taking electrons away from other atoms to form bonds.

When the "ium" atoms meet the "ine" atoms, they are very eager to join. They form ionic bonds where the "ium" atom gives its electrons to the "ine" atom. Table salt (sodium chloride) is an example of an ionic bond between sodium and chlorine.

Other atoms like to **share their electrons** to form **covalent bonds**. Carbon, for example, typically shares its electrons with other atoms, like hydrogen, to form covalent bonds.

Methane is an example of a molecule made from one carbon atom joined to four hydrogen atoms by covalent bonds.

2.5 Summary

- **Atoms** are made up of **protons**, **neutrons**, and **electrons**.
- The number of **protons**, **neutrons**, and **electrons** an **atom** has determines the kind of **atom** it is. **Atoms** with different numbers of **protons**, **neutrons**, and **electrons** are different from each other. **Atoms** with the same numbers of **protons**, **neutrons**, and **electrons** are the same kind of **atom**.
- **Electrons** like to be in a pairs..
- **Bonds** form when **atoms give, take, or share electrons**.
- A **covalent bond** forms when **atoms share electrons**.
- An **ionic bond** forms when **atoms do not share electrons**.

2.6 Vocabulary

Basic Words

atom (AA-tuhm)

boron (BAWR-ahn)

carbon (KAHR-buhn)

chlorine (KLAWR-een)

fluorine (FLAWR-een)

methane (MEH-thayn)

neon (NEE-ahn)

neutron (NOO-trahn)

proton (PROH-tahn)

Challenge Words

beryllium (buh-RIH-lee-uhm)

covalent (koh-VAY-luhnt)

electron (ih-LEHK-trahn)

helium (HEE-lee-uhm)

ionic (iy-AH-nihk)

lithium (LIH-thee-uhm)

nitrogen (NIY-truh-juhn)

oxygen (AHK-sih-juhn)

sodium (SOH-dee-uhm)

CHAPTER 3

Will That Be ICE OR WATER?

3.1 Introduction

Have you ever noticed that water can exist in different forms? If you freeze water, it turns to ice. If you let the ice warm up, it turns to liquid water. If you boil liquid water, it will disappear into the air as water vapor. Each one of these different forms is called a **state**.

Water can exist in three different states: ice, liquid water, and **water vapor**. The general terms for these three states are **solid**, **liquid**, and **gas**. Most substances can exist as a solid, a liquid, or a gas.

3.2 Solids, Liquids, Gases, and Energy

When an atom or molecule exists in different states, it is still the same atom or molecule, but it behaves, looks, and feels different. When water is in the solid form, it can feel cold and be hard to the touch. When it is in the liquid form, it can be flowing and warm. When water is in the vapor (gas) form, you can't usually see it and you can't hold it in your hands. The only difference between these three states is the amount of **energy**.

Recall that energy is something that gives something else the ability to do **work**. Also recall that work, in physics, happens when a **force** moves an object. The only difference between water as a solid, a liquid, or a gas is the amount of energy the individual water molecules have, or the amount of work the molecules are doing. In other words, the only difference between a solid, a liquid, and a gas is how much energy is in the moving molecules.

3.3 Solids

Imagine that you could travel inside an ice cube. You would need to take a warm coat, but imagine that you could make yourself very small — small enough to wander around inside the ice cube. What do you think you would find?

If you could go inside an ice cube, you would find that the water molecules are arranged in a **lattice**. The word lattice refers to the organization of the water molecules in a particular repeating pattern. In an ice cube, although the water molecules are moving, they are moving very slowly and stay organized in a lattice.

Although different solids that are made of different molecules are different from each other, most solids form some kind of lattice. Some solids have highly organized lattice structures and form **crystals**. Minerals like diamonds and rubies are crystals. Other solids, like glass and certain metals, have more loosely organized lattice structures and don't form crystals. However, both highly organized lattice structures and loosely organized lattice structures are solids.

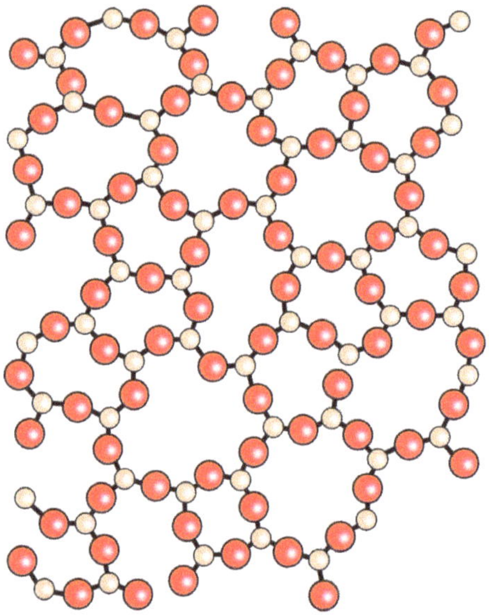

Loosely organized lattice structure

3.4 Liquids

Let's go back to the inside of the ice cube. As you walk around, you might be admiring the wonderful ice lattice. But if someone takes your ice cube and places it at room temperature or outside in the hot sun, your ice cube will start to melt. As the ice cube starts to melt, the water molecules absorb heat energy from the surroundings, and they start to move more quickly. When the molecules are moving fast enough, they will break apart from the ice lattice and actually move closer to each other. If you are still in the ice lattice while

it is melting, you will find yourself getting bumped by fast moving water molecules as the ice cube turns to liquid water.

When we drink water or take a shower, we are interacting with water in the liquid state. When water is in the liquid state, it feels wet and moves around freely. It can spill and be poured, and it does not form a specific shape on its own. You cannot pick it up with your fingers or with a fork or knife. Instead, liquids must be collected in some sort of container like a spoon, cup, or bowl. Liquid water is still the same water it was when it was ice. The molecules just have more energy and are moving more quickly than they do in ice.

3.5 Gases

If you hang out in the liquid water long enough, you will start to see some water molecules **evaporating**, or escaping the surface. A water molecule evaporates because it has absorbed lots of heat energy — so much heat energy that it can't stay with the other water molecules any longer. It must leave and go into the gas state. In the gas state the water molecules move far away from each other so quickly that they move into the air.

When you boil water or are in the shower, you can see steam rising. This steam is water that is in the gaseous state. We usually cannot see or feel or touch gases like we can solids and liquids, but we interact with gases all the time.

When we breathe, we are breathing in a mixture of different gases that are in the atmosphere, and when we blow up a balloon we are filling the balloon with gases from our lungs. When we smell something nice like flowers or stinky like a skunk, it is because gas molecules are making their way to our noses.

Water in the gas state is the same water as when it was in the liquid or solid state. It just has much more energy and the molecules move much more quickly.

3.6 Summary

- Matter exists in 3 major forms: **solids**, **liquids**, and **gases**.
- The only difference between a **solid**, a **liquid**, and a **gas** is the amount of **energy**.
- **Solids** have lower **energy** than either **liquids** or **gases**.
- **Liquids** have more **energy** than **solids**, but less **energy** than **gases**.
- **Gases** have so much **energy** that they move into the air as their molecules quickly move far away from each other.

3.7 Vocabulary

Basic Words

crystal (KRIH-stuhl)

gas (GAAS)

lattice (LAA-tuhs)

liquid (LIH-kwid)

solid (SAH-luhd)

state (STAYT)

Challenge Words

energy (EH-nuhr-jee)

evaporate (ih-VAA-puh-rayt)

CHAPTER 4

Fast and Slow REACTIONS

4.1 Introduction

In Chapter 2 we learned that when atoms combine by giving, taking, or sharing electrons, they make molecules. When atoms switch, join together, or break apart by giving, sharing, or taking electrons, a **chemical reaction** occurs.

Sometimes these chemical reactions happen quickly, and sometimes they happen slowly. Whether a reaction occurs quickly or slowly depends on how close atoms are to each other and how fast they are moving.

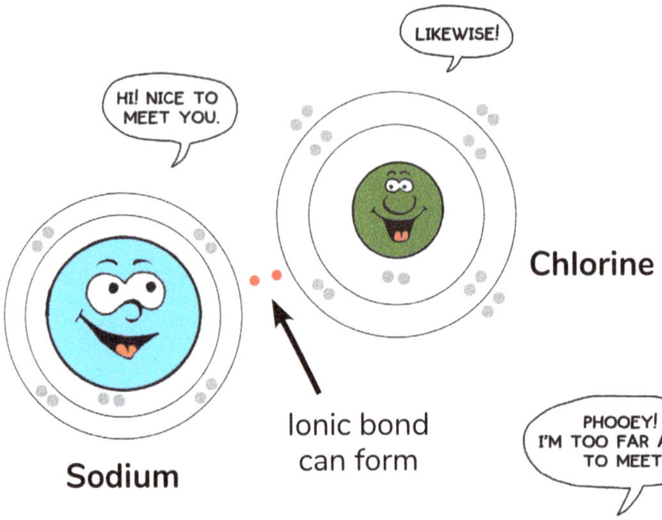

In order for a chemical reaction to occur, two atoms or molecules must be close enough together to meet. If a sodium atom and a chlorine atom are close enough to each other that they can bump, they will form a table salt molecule (sodium chloride).

Also, in order for a reaction to occur, two atoms or molecules must be moving fast enough to bump into each other. If they are moving too slowly, they won't have a chance to bump into each other.

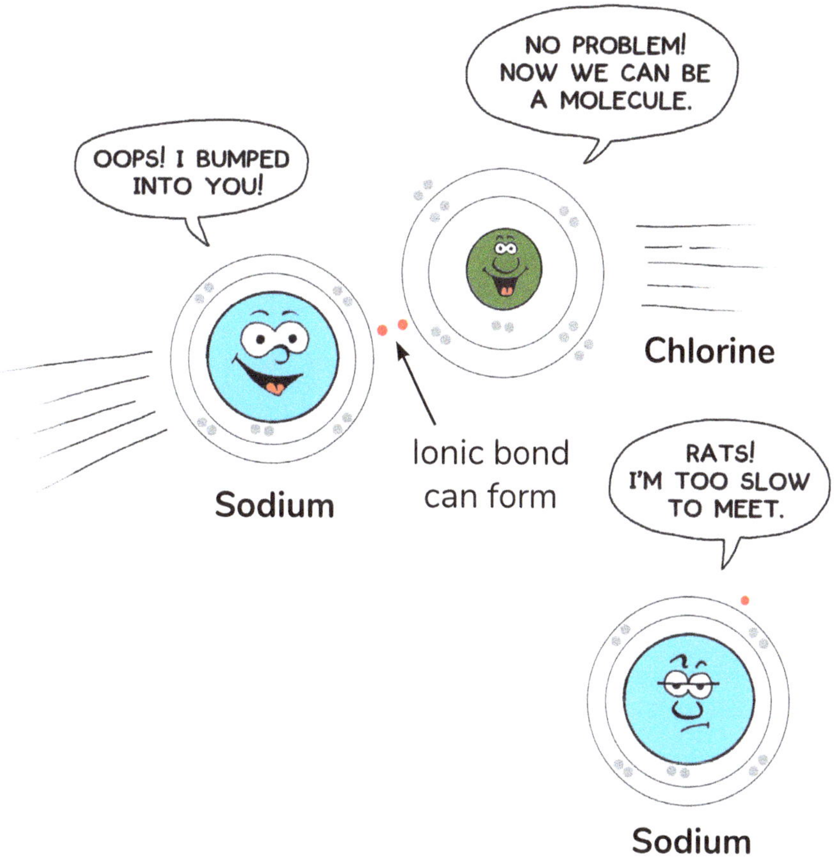

4.2 Fast Reactions

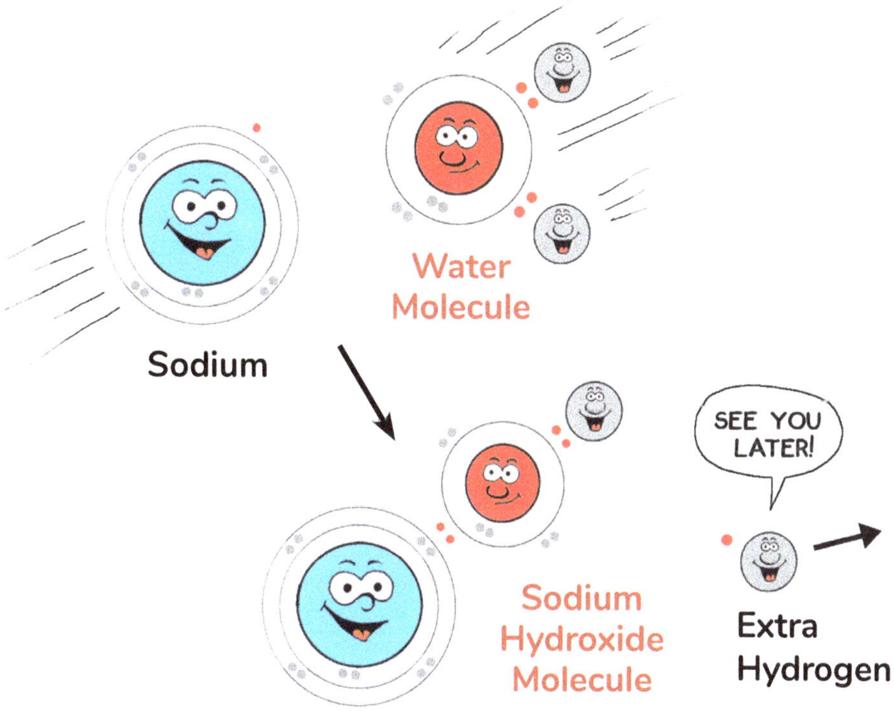

Some chemical reactions occur very quickly. For example, if a sodium atom and a water molecule are placed in a container together, they will run to meet each other and will very quickly join to form a **sodium hydroxide** molecule.

In fact, sodium atoms and water molecules are so eager to meet that when they get close enough to each other, they explode!

Sodium atoms and water molecules reacting

4.3 Slow Reactions

Other reactions occur very slowly. A shovel is made of iron, and when it is left outdoors, the oxygen and water in the air can react with the iron to form rust, which is called **iron oxide**. This is a very slow reaction, and it depends on how much oxygen and water are available in the air. If there is little water in the air, the reaction goes very slowly. However, if there is more water in the air, the reaction will go more quickly. Also, salt can affect how quickly iron rusts. In humid climates near sea water, iron on cans, shovels, and even buildings can rust more quickly than in dry climates.

4.4 Ways To Speed Up and Slow Down Reactions

People often need slow reactions to occur more quickly than they normally would and fast reactions to occur more slowly than they normally would. If we could not speed up reactions, then many processes that we rely on would take too long, and if we could not slow down reactions, some processes would go too fast. There are a number of different ways to "help" a reaction go more quickly or more slowly.

Heat

When you apply **heat**, molecules have more energy to move around, so they move faster. The faster molecules move, the more likely they are to meet other molecules and undergo a reaction. When we cook foods we are speeding up reactions.

If you take heat away, molecules slow down and there is less chance of a reaction. When you don't want a reaction to occur, or you want to slow it down, you lower the temperature to make the molecules move more slowly. This is why you put food in the refrigerator or freezer.

Some reactions can occur so quickly that it is important to try to prevent them from happening. This is why it is important to keep flammable liquids and gases away from open flames and sources of heat.

Concentration

Another way to increase or decrease the likelihood of two molecules meeting to form a reaction is to increase or decrease the **concentration.** Concentration is the number of molecules that are in a particular space.

When there are more molecules in a space, that space becomes more crowded, and molecules are more likely to bump into one another. The more they bump into one another, the more likely they are to undergo a reaction. When there are fewer molecules in an area, the molecules are less likely to bump into one another and are less likely to undergo a reaction.

4.5 Summary

- A **chemical reaction** occurs when atoms combine to make a molecule.
- In order for a **chemical reaction** between atoms or molecules to occur, the atoms or molecules must be moving and they must meet.
- Some **chemical reactions** occur quickly and others occur slowly.
- A **chemical reaction** can be made to occur more quickly by adding **heat** or increasing the **concentration** of molecules (making the molecules more crowded).
- To slow down a **chemical reaction**, you can decrease the temperature or the **concentration** (make the molecules less crowded).

4.6 Vocabulary

Basic Words

iron oxide (IY-uhrn AHK-siyd)

Challenge Words

chemical reaction (KEH-mih-kuhl ree-AAK-shuhn)

concentration (kahn-suhn-TRAY-shun)

sodium hydroxide (SOH-dee-uhm hiy-DRAHK-siyd)

CHAPTER 5

What Is HEAT?

5.1 Introduction

As we saw in the last two chapters, heat is very important for helping chemical reactions take place and for helping solids turn to liquids and gases. We learned that heat can help reactions occur more quickly and that taking heat away can help reactions occur more slowly.

In this chapter we will study how heat is absorbed and released in different chemical reactions and how heat changes atoms and molecules from one state to another without a chemical reaction. The study of heat is called **thermodynamics**. The word thermodynamics comes from the Greek words *therme* which means "heat" and *dynamikos* which means "power."

5.2 Energy for Making and Breaking Bonds

Energy is usually released when chemical bonds in molecules are broken. Humans and animals get energy by breaking down chemical bonds in the food they eat. They use this energy for necessary processes, such as making new bonds and new cells.

When we eat, we start by using our teeth to chew the food into little pieces. After swallowing the food, the acids in our stomach break the bonds that hold the chemicals in the food together. The breaking of these bonds releases heat. Heat created in this way is a form of energy that the body can use for performing different processes, such as building muscle and repairing wounds. When you are young and growing or if you are sick or injured, it is especially important to eat nutritious food because your body needs extra energy to heal itself.

Heat is the form of energy that humans and other animals use to stay alive. You feel warm after eating or exercising because you are breaking a lot of chemical bonds and this is releasing heat. Your body does much of its cell and muscle building when you are sleeping. In order to occur, these processes require heat energy. This is why you may feel colder while sleeping and may want to use a blanket.

5.3 Where Does Heat Come From and Where Does It Go?

When some chemical reactions take place, they release heat. Other chemical reactions cause heat to be absorbed. Reactions that release heat energy are called **exothermic**, and those that absorb heat energy are called **endothermic**.

Exothermic and endothermic are words that are derived from Greek. *Exo* means "outside" and *endo* means "inside." This is easy to remember because you go outside when you exit, and you come inside when you enter. And, as was mentioned before, the Greek word *therme* means "heat."

When an **exothermic** reaction occurs, heat is released (goes out) from the chemicals that are reacting. You can often feel this heat. For example, a candle burning is an exothermic reaction.

When a candle burns, you can feel the heat that comes from the flame. The heat you feel comes from the energy already present in the bonds of the molecules. The burning of these molecules is a chemical reaction. As the bonds of the molecules are broken, heat energy is released and transferred to the surroundings.

An **endothermic** reaction occurs when you bake a cake. You put the batter in the oven and bake it at a certain temperature. While the cake is baking, the chemicals in the eggs, flour, salt, sugar, and other ingredients are rearranging themselves to form new bonds. The formation of new bonds is endothermic, meaning it uses heat energy. This is why, without an oven to heat the batter, the cake will not form. The heat produced by the oven is not disappearing. Instead, it is being absorbed by the cake and is being used to build new bonds.

5.4 Heat Laws

Much of science is based on making observations about our surroundings, particularly those in nature. In science there are different laws that are observed to always take place, without exception. These laws do not necessarily explain what, why, or how something occurs, just *that* it occurs. Explaining the what, why, and how is left to the observer (that's you!) to think about and try to explain through the use of experiments.

The law known as the **First Law of Thermodynamics** states that **energy cannot be created or destroyed**. When you feel something get hot, heat is not being created, but merely **transferred** to the part of your body that feels the heat.

The First Law of Thermodynamics applies to chemical reactions and to changes of state.

As we saw in Section 5.3, when you bake a cake, the heat from the oven does not disappear. Instead, it is transferred to molecules inside the cake through chemical reactions.

When you heat an ice cube, the heat that is being applied to the ice cube does not disappear. Instead, it is absorbed by the ice cube. This is why the temperature of the ice cube starts to increase as it melts to liquid water and the temperature of the liquid water increases until it boils and turns into gas. In these changes of state, the heat is being absorbed and transferred throughout the water molecules but there is no chemical reaction taking place.

5.5 Summary

- **Thermodynamics** is the study of heat.
- Chemical reactions that release heat are called **exothermic**. Chemical reactions that absorb heat are called **endothermic**.
- The **First Law of Thermodynamics** states that **energy is not created or destroyed** but it can be transferred between substances, usually in the form of heat.
- Heat from chemical reactions comes from energy already present in the bonds of molecules. Heat energy is released as the bonds are broken.

5.6 Vocabulary

Challenge Words

endothermic (ehn-duh-THUR-mihk)

exothermic (ehk-soh-THUR-mihk)

thermodynamics (thuhr-moh-diy-NAA-mihks)

CHAPTER 6

What Are ANIMALS?

6.1 Introduction

When you think of an animal, what do you think of? Do you think of a cat? A dog? A frog? A spider?

When many people think of an animal, they think of a pet or an animal they've seen at the zoo. They might think of creatures that have a head, eyes, nose, mouth, legs, fur, and possibly a tail.

But some animals are very different from a dog, a cat, or an elephant at the zoo. Some animals, like fish, have scales. Some animals have wings, like dragonflies, and some animals look like plants but aren't plants. Sponges can look like plants but are really animals.

6.2 What Is an Animal?

There are many different features that distinguish animals from other living things, but there are a few basic features that all animals share.

1. All animals have many cells that work together. With the exception of sponges, all animals have different parts of their body doing different things with different **tissues** and sometimes with different **organs**.

A **tissue** is a group of cells that perform a particular function. Animals, even worms, have muscle tissues, brain tissues, and tissues for circulation. These different tissues are made of many different cells.

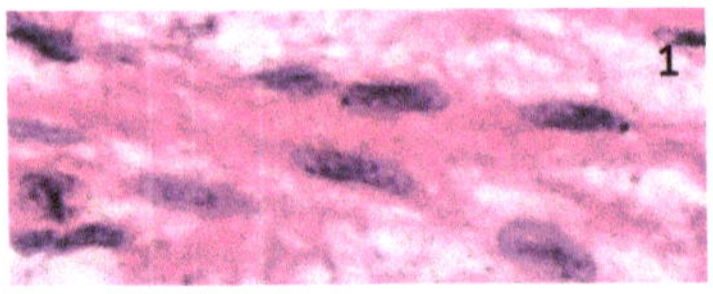

Muscle cells

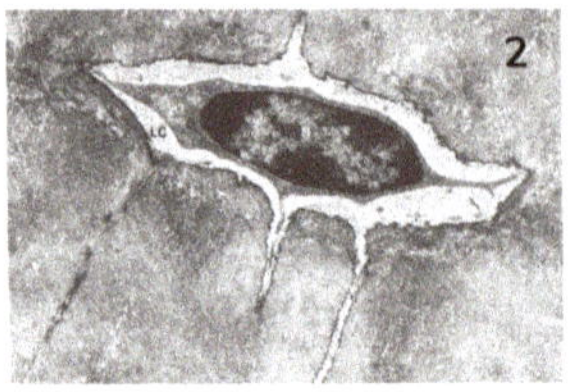

Bone cell

Skin cells

Tissue Cells

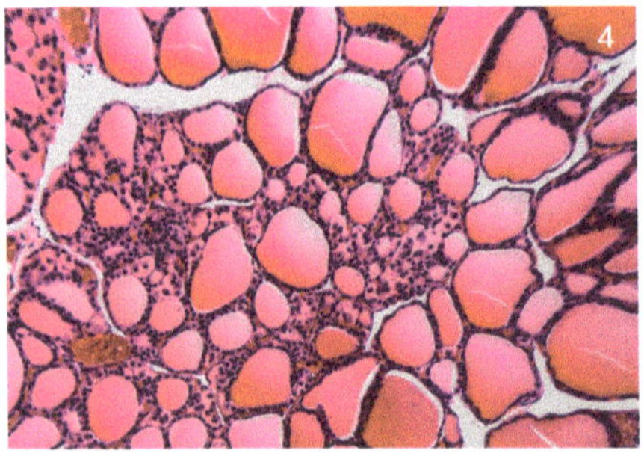

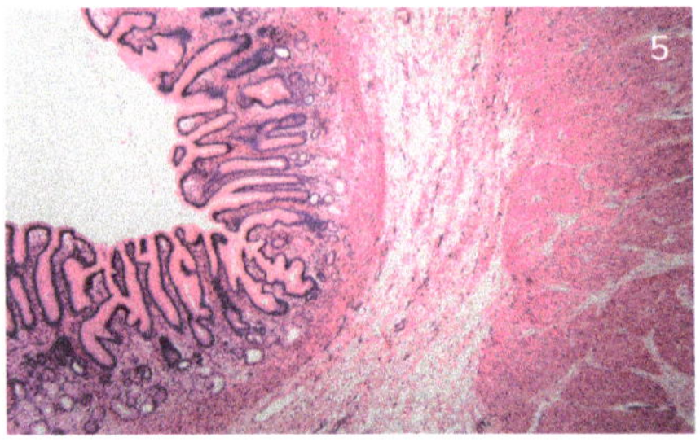

Tissues

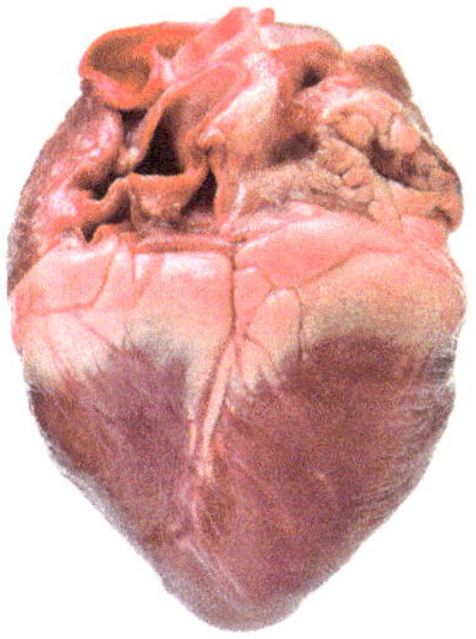

Organs are made of different tissues that together perform a particular function. The heart is an organ that pumps blood, and the lungs are an organ that captures oxygen for the body to use.

2. Animals do not make their own food but instead eat plants or other animals. Some animals eat only plants. These animals are called **herbivores**.

Some animals eat only other animals. These animals are called **carnivores**.

Many animals, including humans, eat both plants and other animals and are called **omnivores**.

3. At some point in their life cycle all animals move independently. Some animals move by walking, some by crawling, and some animals fly. Some animals, like the sponges, move only during the first part of their life cycle, but most animals move all the time.

6.3 Animal Cells

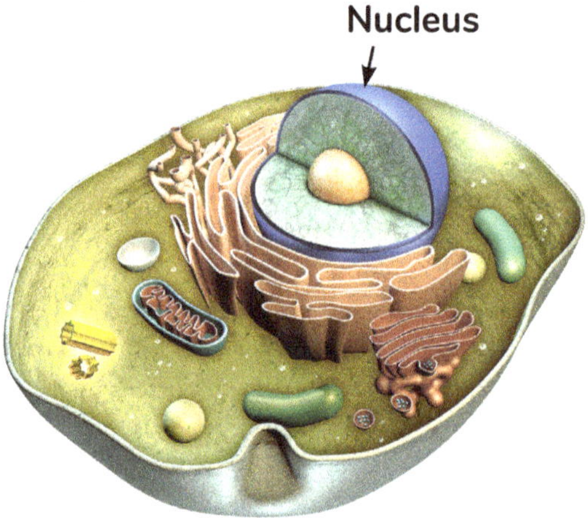

All animals are in the **domain Eukarya**, and all animals have animal cells. Animal cells are different from the cells of **bacteria** and **archaea** because animal cells have a **nucleus**.

Animals cells are not rigid and are not used by animals for structure like they are in plants. Some animal bodies, like those of sea jellies, are very soft and lack a rigid structure. Other animals, like crabs and insects, have cells that form hard outer tissues. Animals such as cats, dogs, and humans have cells that form bone for structure.

6.4 Animal Phyla

Because there are so many different kinds of animals, scientists are still arguing about exactly how to divide and organize all the animals into groups. But they begin by dividing the animal kingdom into smaller groups called **phyla**. There are anywhere from 35 to 42 different phyla, depending on the taxonomic system being used. However, most animals fit into just nine phyla.

Eight of the nine phyla are animals with no backbone. A backbone is a series of interconnected bones that run down the back of some animals. Animals with no backbone include: **Porifera** (sponges), **Cnidaria** (sea jellies, coral, etc.), **Platyhelminthes** (flatworms), **Nematoda** (roundworms), **Annelida** (segmented worms), **Mollusca** (snails, octopuses, clams, etc.), **Echinodermata** (sea stars, sand dollars, sea urchins, etc.), **Arthropoda** (insects, spiders, lobsters, etc.).

The ninth phylum is **Chordata** (bird, cat, fish, human, etc.) which includes the animals that do have backbones.

Porifera
(sponges)

Cnidaria
(jellyfish, coral, etc.)

Platyhelminthes
(flatworms)

Nematoda
(roundworms)

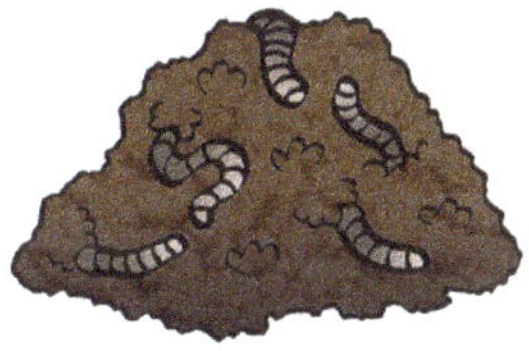

Annelida
(segmented worms)

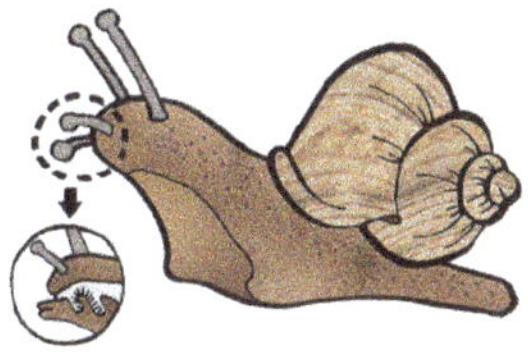

Mollusca
(snails, octopuses, etc.)

Echinodermata
(sea stars, sea urchins, etc.)

Arthropoda
(spiders, lobsters, etc.)

Chordata
(humans, fish, birds, etc.)

Domains, kingdoms, and phyla – putting it all together

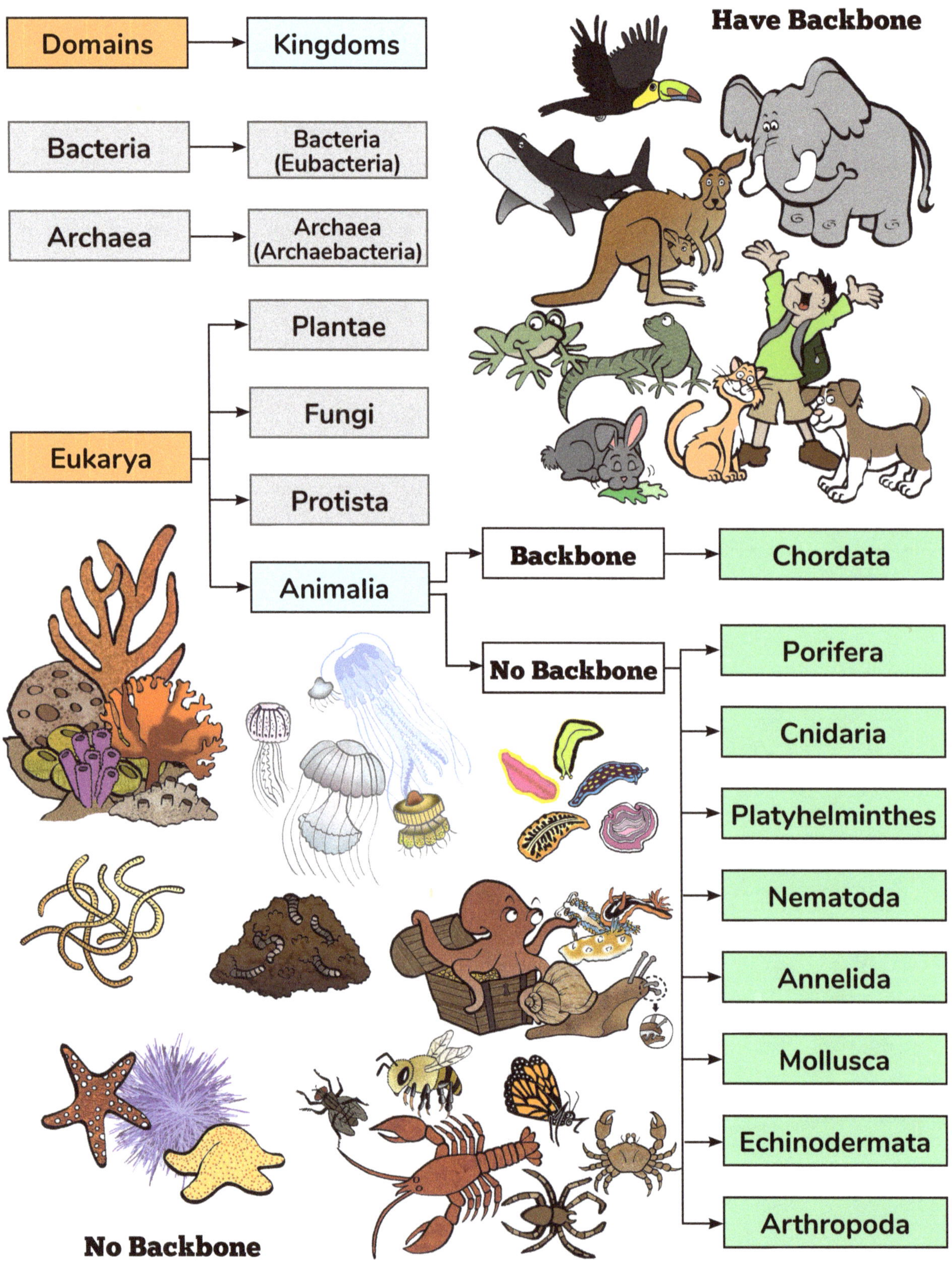

6.5 Summary

- All animals have many cells, eat other plants or animals, and move independently at some point in their life cycle.
- Some animals are **herbivores** and eat only plants.
- Some animals are **carnivores** and eat only other animals.
- Some animals are **omnivores** and eat both plants and animals.
- Animals have **animal cells** that have a **nucleus**.
- Even though there are between 35 and 42 **phyla**, most animals fit into 9 major **phyla**.

6.6 Vocabulary

Basic Words

domain (doh-MAYN)

organ (AWR-guhn)

phylum (FIY-luhm) **phyla** *[plural]* (FIY-luh)

tissue (TIH-shoo)

Challenge Words

Annelida (aa-NEHL-uh-duh)

archaea (aar-KEE-uh)

Arthropoda (aar-THRAH-puh-duh)

bacteria (baak-TIH-ree-uh)

carnivore (KAHR-nuh-vawr)

Chordata (kawr-DAH-tuh)

Cnidaria (niy-DAAR-ee-uh)

Echinodermata (ih-kye-nuh-DEHR-muh-tuh)

Eukarya (yoo-KAA-ree-uh)

herbivore (EHR-buh-vawr)

Mollusca (muh-LUHS-kuh)

Nematoda (nehm-uh-TOH-duh)

nucleus (NOO-klee-uhs)

omnivore (AHM-nuh-vawr)

Platyhelminthes (plaa-tih-hehl-MIHN-theez)

Porifera (puh-RIH-fuh-ruh)

CHAPTER 7

Squishy, Spiny, Slimy ANIMALS

7.1 Introduction

There is a lot more to the animal kingdom **Animalia** than just the animals we might see every day, like cats and dogs.

Cats, dogs, birds, elephants, and humans all have a **backbone**. A backbone is a series of interconnected bones called **vertebrae** that run down the back of some animals. You have a backbone that runs from the base of your head to your tailbone. Animals that have a backbone are called **vertebrates**.

But did you know that about 95% of the animals in the world do not have a backbone? This means that most of the animals on the planet are not like cats and dogs and people.

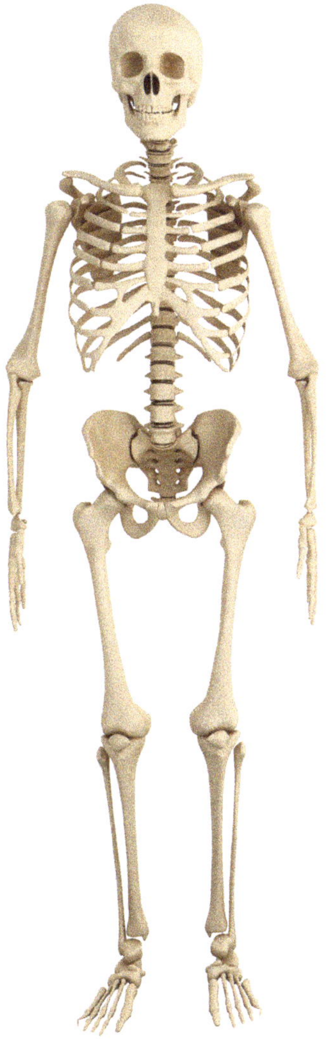

Human skeleton

In this chapter we will take a look at some of the animals that don't have a backbone. Animals in this group are called **invertebrates**.

Invertebrates—95%

Vertebrates—5%

7.2 Sponges

A **sponge** is a simple animal that is in the phylum **Porifera**. Although sponges look like plants, they are actually animals. Sponges are made of animal cells, are multicellular, eat food, and spend at least some part of their life being mobile. Sponges live in water, and most sponges live in the ocean. Some of the sponge cells have whiplike **flagella** that wave and create a current of water that flows through the sponge body. The water contains bacteria and other tiny bits of food for the sponge to eat.

Sponges can be almost any shape. Some sponges are small and round, and some are tube-shaped, like a straw. Most sponges are smaller than a few inches, but a few, like the barrel sponge, grow so large that a diver could fit inside.

Sponges don't have a backbone, but they do have a skeleton made of a mesh of protein called **spongin**. If you take a sea sponge out of the water and let it dry, the spongin remains. This is what people use for cleaning or scrubbing. Spongin is soft, but the mesh structure makes it very tough.

7.3 Sea Jellies (Jellyfish)

Sea jellies are a type of invertebrate in the phylum **Cnidaria**. Sea jellies live all over the world, in shallow and deep water. Some are huge, growing bigger across than an adult is tall. Some have **tentacles** (long, stringy appendages) that are longer than an American football field. Some sea jellies are smaller than your fingers.

Although people often call them jellyfish, sea jellies are not fish and do not have a backbone. When they have washed up on the beach, sea jellies are shapeless like Jell-O or gelatin. Because a sea jelly's body is soft and without structure, gravity deforms its shape if it ends up on land. But in the ocean, a sea jelly can move effortlessly through water by pulling and pushing the water through its body.

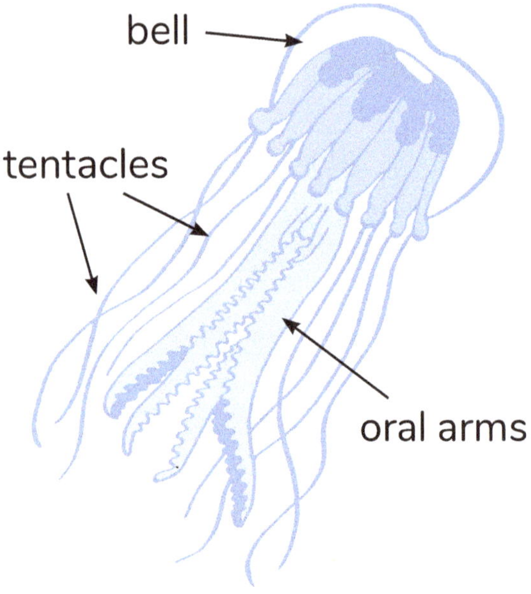

The sea jelly's body is called its **umbrella**, or **bell.** Many sea jellies have long, thin tentacles attached to the body. The tentacles hang down from the body and are used to catch food. The sea jelly's mouth is in the center of its bell. As the sea jelly floats up and down in the water, food is pushed toward its mouth. **Oral arms** found around the mouth sweep food into the mouth. Sea jellies eat both large and small creatures from the sea, including tiny **plankton** and large fish.

7.4 Worms

All worms have long tubular or flat bodies, no backbone, and no legs. There are actually many different types of worms, and all worms are grouped into three main categories: **flatworms**, **roundworms**, and **segmented worms**.

Flatworms are in the phylum **Platyhelminthes**. Flatworms are long, skinny and flat. Some flatworms are very colorful and live in the ocean. Many of these flatworms live on coral reefs. They eat plants or animals that don't move much, such as sponges. Some ocean flatworms have little hairs on their underside that help them glide over the sand on the ocean floor.

Some flatworms live on land or inside other animals, like cows, and even humans. We call these flatworms **parasites** because they live inside other organisms. These parasites get their food from other animals' bodies, usually without giving anything beneficial back to the animal. They can make animals sick and sometimes can cause the animal to die.

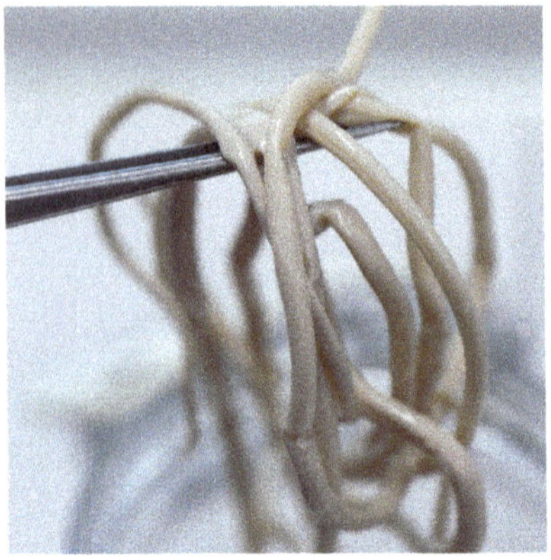

Roundworms are in the phylum **Nematoda**. Roundworms are small, smooth, and round. Roundworms live in the ocean and on land. There are close to 30,000 different types of roundworms, and many of those are parasites. Roundworms often infect pigs. Because of these roundworms, humans can become sick if they eat pork that has not been thoroughly cooked.

Segmented worms are in the phylum **Annelida**. Like roundworms, segmented worms are also round but have segmented bodies instead of smooth ones. **Earthworms** are a type of segmented worm. If you look closely at an earthworm, you can see what appear to be little bands or rings circling it. These bands occur between the **segments**, or sections, of the earthworm's body. There are over 22,000 different kinds of segmented worms!

Some segmented worms, such as earthworms, are good for gardens. They eat decaying plants and tiny organisms in the soil and return nutrients to the soil in a form that can be used by plants. Earthworms also create little tunnels in the soil, allowing air, water, and nutrients to go deep into the soil. To make a tunnel, the earthworm squeezes its muscles and then relaxes them, making its body stretch and move forward. It eats dirt as it moves along.

7.5 Snails and Octopuses

Even though **snails** and **octopuses** look very different, they are both in a group of invertebrates called **mollusks** in the phylum **Mollusca**. Although mollusks look very different from one another, they all share a similar body plan. All mollusks have soft bodies, use one or more muscular feet to move, and have a membrane surrounding their internal organs that can sometimes form a hard shell.

The basic body plan of mollusks can be easily observed in snails. **Snails** have a large muscular foot below their soft body. They use this foot to move. They also have a hard shell.

Snails are found all over the world and are not often welcome in the garden. Snails and their cousins, **slugs**, eat some of our favorite garden plants. Inside a snail's mouth is an armored, spiky "tongue" called a **radula**. The snail uses the radula to cut plants and scrape up **algae** to eat. Snails, like worms, live in areas that are moist and not too cold or hot. Most snails are small but some sea snails can grow to be almost 1 meter (3 feet) long.

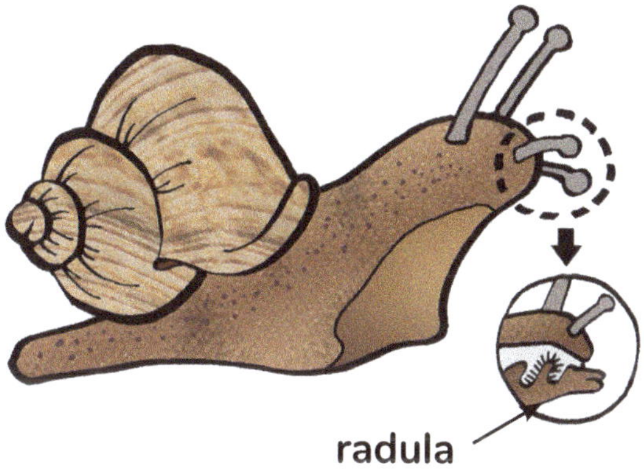

Saltwater snails are often referred to as **sea slugs** because they don't have an outer shell. These snails live in shallow ocean waters and eat plants, sponges, corals, or other slow animals. Many saltwater snails are small, but some are big enough to stretch from your hand to your elbow.

Octopuses don't look much like snails, but they are related. Octopuses don't have a hard outer shell, and many have eight "feet" called ams. The word octopus comes from the Greek words *okto* which means "eight" and *podos* which means "foot."

Some octopuses are only a few centimeters (inches) long, but the giant octopus can grow up to 9 meters (30 feet) long. Octopuses live on the sea floor in both shallow and deep ocean water. Octopuses swim by sucking water in, then squirting it out behind them. They eat worms, crabs, and snails, and some use the suckers on their arms to catch fish.

Octopuses spend time hunting and hiding. Because they have no shells, they can squeeze themselves in small spaces between rocks to hide. Many octopuses can change their colors and markings to blend in with their surroundings, making it difficult for big fish to notice them.

Octopuses are very smart. Scientists have taught octopuses to find their way through mazes, and with enough time, an octopus can open a jar to retrieve food. Scuba divers have found octopuses to be very curious animals.

7.6 Sea Stars, Sand Dollars, Sea Urchins

Sea stars, **sand dollars**, and **sea urchins** are part of a large group of invertebrates called **echinoderms**. The word echinoderm means spiny skin, and all of the animals in this group have rough or prickly outer skin. They are in the phylum **Echinodermata**.

Sea stars are often called **starfish** although they are not related to true fish. Some sea stars are as small as your hand, and some can grow to be 1 meter (3.3 feet) across. Most sea stars are flat, spiny, and have five arms that stretch out in all directions from the center. Some have more than five arms. The mouth is on the underside in the center of the body.

Sea stars cannot raise themselves very far off the rocks or use their arms like people use legs. If you turn a sea star over, you will see lots of little tube-shaped feet all the way down each arm. These little feet are used to hold on to rocks and to creep slowly along the ocean floor. Many sea stars can also use the suckers on their feet to hold on to a mussel or clam shell and pry it open so they can eat the animal inside. Sea stars also eat snails, sea urchins, and other small ocean animals.

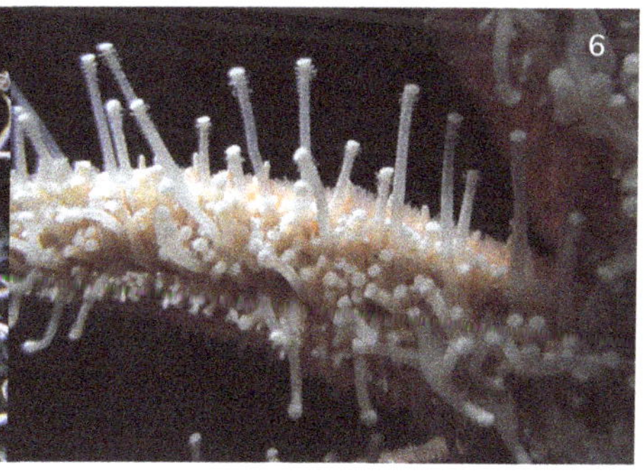

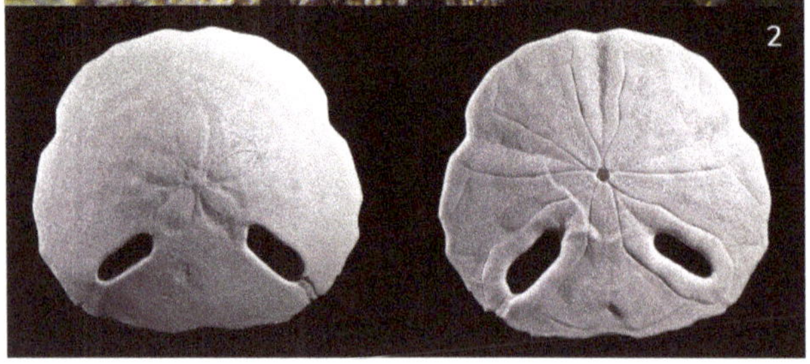

Have you ever seen a **sand dollar**? If you have seen one used as a bathroom decoration, you have actually seen the skeleton of a sand dollar. Did you know that there are five "petals" outlined in the pattern on the top of a sand dollar's skeleton? The sand dollar is a close relative of the sea star and the sea urchin.

A sand dollar has a flat, disk-shaped body no bigger than the palm of your hand. Like the sea star, it is covered with very short spines, has lots of tube feet, and has a mouth on its underside. Sand dollars live close to shore and hide under the sand or mud when a hungry sea star comes creeping by or when the water gets rough. The sand dollar uses its spines to bury itself, to catch food and bring the food to its mouth, and to creep over the sand. To catch tiny larvae or bits of food floating in the water, the sand dollar can stand itself on end with one edge buried in the sand.

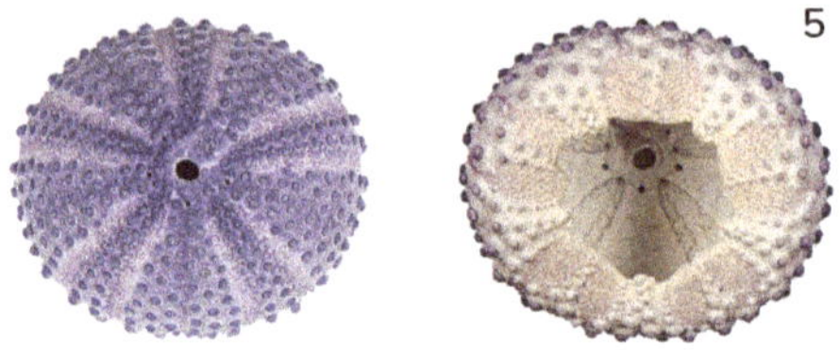

The **sea urchin** is the spiniest underwater creature of all. It looks like a little round pincushion. The size of most sea urchins is 3 to 10 cm (1 to 4 in). A sea urchin can move its blunt spines, like sand dollars do. Sea urchins use their spines to protect themselves and to move around. They have five rows of long, hair-like tube feet that are just a bit longer than their spines.

These tube feet hold on tightly to rocks so the sea urchin can eat the algae growing on the rocks. Like sea stars and sand dollars, sea urchins have a mouth on the underside. Like sand dollars, they also have a "five petal" pattern on their skeleton. Most sea urchins can be found close to shore in oceans all over the world. Some live in the deep ocean.

Echinoderms are interesting to observe but can be harmed if touched or captured. Some have spines that can puncture skin.

7.7 Summary

- Animals without a **backbone** are called **invertebrates**.
- About 95% of the animals in the world **do not** have a **backbone**.
- **Sponges** eat by taking food particles out of water that moves through their body.
- **Sea jellies** are soft and have **oral arms** to move food into the mouth.
- **Worms** have long tubular or flat bodies. There are three kinds of worms: **flatworms**, **roundworms**, and **segmented worms**.
- **Snails** and **octopuses** are **mollusks**.
- **Sea stars**, **sand dollars**, and **sea urchins** are **echinoderms** and have spiny bodies and tube feet.

7.8 Vocabulary

Basic Words

alga (AAL-guh) **algae** *[plural]* (AAL-jee)

mollusk (MAH-luhsk)

segment (SEHG-muhnt)

sponge (SPUHNJ)

spongin (SPUHN-jihn)

Challenge Words

Animalia (aan-uh-MAHL-yuh)

echinoderm (ih-KIY-nuh-duhrm)

flagellum (fluh-JEH-luhm) **flagella** *[plural]* (fluh-JEH-luh)

invertebrate (ihn-VUHR-tuh-bruht)

octopus (AHK-tuh-puus)

radula (RAA-juh-luh)

tentacle (TEHN-tuh-kuhl)

umbrella (uhm-BREH-luh)

vertebra (VUHR-tuh-bruh) **vertebrae** *[plural]* (VUHR-tuh-bray)

vertebrate (VUHR-tuh-bruht)

CHAPTER 8

Stinging, Crawling, Squirming ANIMALS

8.1 Introduction

In the last chapter we looked at several different groups of invertebrates — animals that don't have a backbone. In this chapter we'll take a look at a very large group of invertebrates. The animals in this group are called **arthropods** and include spiders, lobsters, crabs, millipedes, flies, bees, and ants. That's a lot of animals!

No matter where in the world you live, if you go outside, walk down the block, go to the park, or get on a boat and sail the ocean, you are likely to run into at least one member of the arthropod group. Arthropods are the most abundant group of animals on the planet. In fact, some scientists estimate that 3 out of 4 animals on the planet are arthropods.

8.2 Basic Body Plan

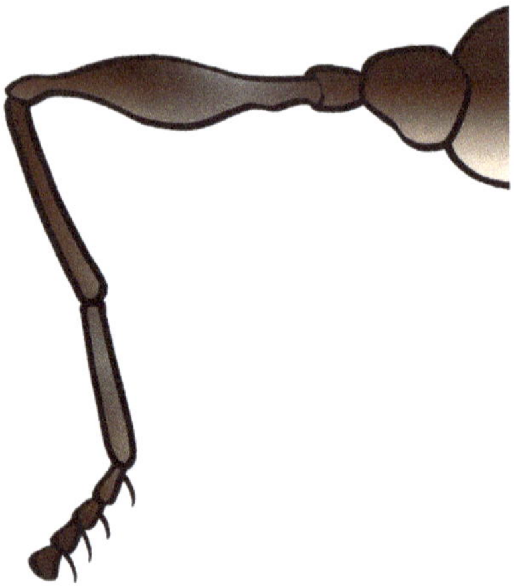

The word **arthropod** comes from the Greek words *arthron* meaning "a joint" and *podos* meaning "foot," so arthropod means "jointed foot." If you look closely at a crab or an ant or a beetle, you can see that their legs ("feet") are segmented, or jointed. All of the animals in this group have segmented, or jointed, legs.

The whole body of an arthropod is covered with a tough outer shell that acts like a suit of armor, protecting and supporting the soft inner body. This tough shell is called an **exoskeleton** and is made of a protein called **chitin**. Chitin is a type of polymer made with **glucose molecules** and **nitrogen**.

In some arthropods, like caterpillars, the chitin is soft and leathery. But in other arthropods, like lobsters and crabs, the chitin combines with other elements, such as calcium, to make shells that are much harder.

Arthropod bodies are divided into segments. **Insect bodies** are divided into three segments called the **head**, **thorax**, and **abdomen**. **Spiders** and **lobsters** are divided into two main segments — a head and thorax fused together into the **cephalothorax**, and the abdomen. Arthropod bodies are segmented so they can move. You can imagine how impossible it would be to move with a full coat of armor if it were all one piece. Like a medieval knight's armor, arthropod bodies need to be divided.

8.3 Insects

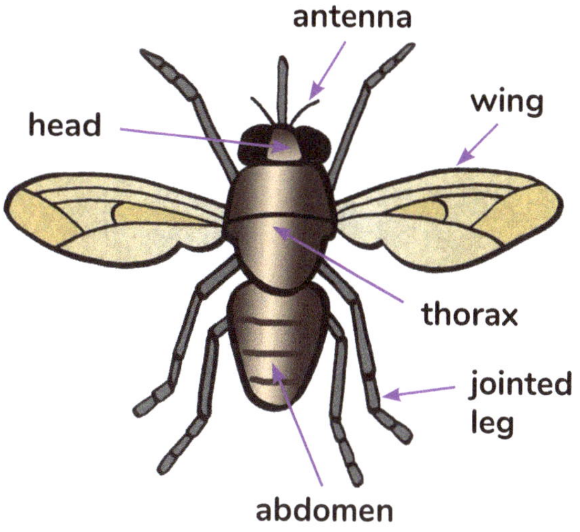

If you go outside, you are likely to find at least one **insect**. Insects live almost everywhere, in all kinds of climates, and they make up the largest group of arthropods. All insects have three pairs of **jointed legs**, **compound eyes**, one pair of **antennae**, and a set of **mandibles** (mouthparts), and most have a pair of wings attached to the thorax.

Insects are grouped according to their wing type. Some insects, like **beetles**, have **sheathed wings,** meaning that their wings are tucked under an exoskeleton **sheath**, or covering. If you look carefully at beetles, you'll see that they can tuck their wings behind their thick exoskeleton sheath. When they are ready to fly, they pull their wings out and take off.

Other insects, like bees, have **transparent** wings that are not protected by an exoskeleton sheath. **Flies**, **mosquitoes**, **bees**, **wasps**, and **ants** that fly all have **unsheathed wings**.

Although many insects have wings, not all insects do. **Fleas** are insects that have small bodies and a pair of strong jumping legs. Fleas don't get around by flying with wings but by jumping from place to place. Fleas are always looking for a warm body to chew on.

BUTTERFLIES

Butterflies are insects that belong to the group of insects called **Lepidoptera**. All of the insects in this group have scales on their wings. If you look closely at a butterfly, you might see tiny scales on its wings.

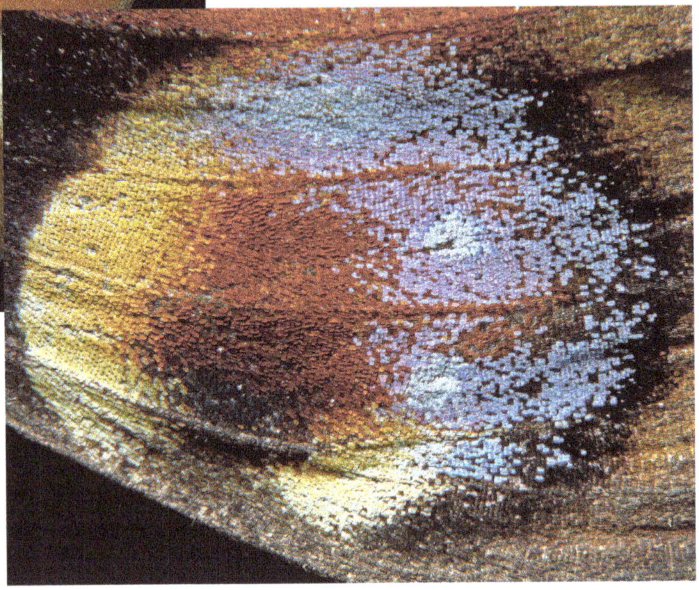

The Beginning: Eggs

Butterflies, like other insects, begin life as an egg. Butterfly eggs are usually very small. Some are fuzzy, and some have ridges. Some are white and round, and some are thin and orange. Because butterfly eggs are so small, they can be hard to see.

The mother butterfly tries to find a safe place to lay her eggs. She looks for a place where the eggs will have food when they hatch and that will be warm and dry and protected from other bugs. The mother butterfly will often glue the eggs to the back of a green leaf. The leaf protects the eggs and can also be used for food when the eggs hatch.

The Middle: Caterpillar

When an egg hatches, a caterpillar is born. The caterpillar comes out of the egg, and the very first thing it does is eat. In fact, eating is about all the caterpillar does. Most caterpillars eat plants, but a few will eat other insects. The main job of the caterpillar is to eat enough food so when the time is right, it will be ready to become a butterfly.

Caterpillars come in all different colors, sizes, and textures. Some are green, and some are brown. Some are smooth, and some are furry. Some are very small, and some grow big and fat. Some even have lots of stripes in pretty colors. There are many different kinds of caterpillars that turn into many different types of butterflies.

As the caterpillar grows, its skin stays the same size. Several times while the caterpillar is growing, it will **molt**, or shed its too small skin, and form new skin.

The Change: Chrysalis

Once the caterpillar has grown as big as it needs to grow, it finds a safe place to stay while it turns into a butterfly. The caterpillar will look for a place like the underside of a leaf, ledge, or tree limb where it can attach itself with a small button of silk. Then it sheds its skin for the last time. The new skin becomes the hard shell of the **chrysalis**.

A chrysalis is like a cozy house that keeps the caterpillar warm and dry. Some chrysalises cover the whole caterpillar and are very thick and tough. But other chrysalises are thinner, and you can see the caterpillar underneath.

Inside the chrysalis the caterpillar changes into a butterfly. The body of the caterpillar starts to change, and wings begin to grow. The long feet of the butterfly grow, and tentacles begin to show.

When the caterpillar has completely changed, or **transformed**, it is ready to emerge from the chrysalis as a butterfly. The process of transforming from one form (the caterpillar) to a different form (the butterfly) is called **metamorphosis**.

The End: Adult Butterfly

When the butterfly is ready to come out, a hole forms in the chrysalis. The butterfly must wriggle through the hole to get out. It looks like the butterfly struggles, but the tiny hole actually helps the wings of the butterfly to be ready for flying.

Once the butterfly has emerged from the chrysalis, it takes a moment to spread its new wings. The wings continue to expand as water from the body is pumped into them. In a few hours the butterfly is able to spread its wings and fly away.

Most adult butterflies eat nectar from flowers, and they sometimes even travel long distances. When the adult female butterfly is ready, she will lay eggs, starting the life cycle for a new butterfly.

8.4 Spiders

Although **spiders** look like insects and can be confused with insects, spiders are not in the same group as insects. Spiders are in a group called **arachnids**. Like insects, spiders have segmented bodies and a chitin exoskeleton, but unlike insects, spiders have **four pairs of legs**, no antennae, and the head and thorax are fused together into the cephalothorax. This means that spiders have only **two segments**: the **cephalothorax** and the **abdomen**, instead of three segments like insects.

Spiders also have two sets of specialized parts called **chelicerae** and **pedipalps**. The chelicerae are fang-like appendages that can deliver poison and are used for killing and eating prey. Just behind the chelicerae are the pedipalps, appendages that are used to grab and position the prey for the chelicerae.

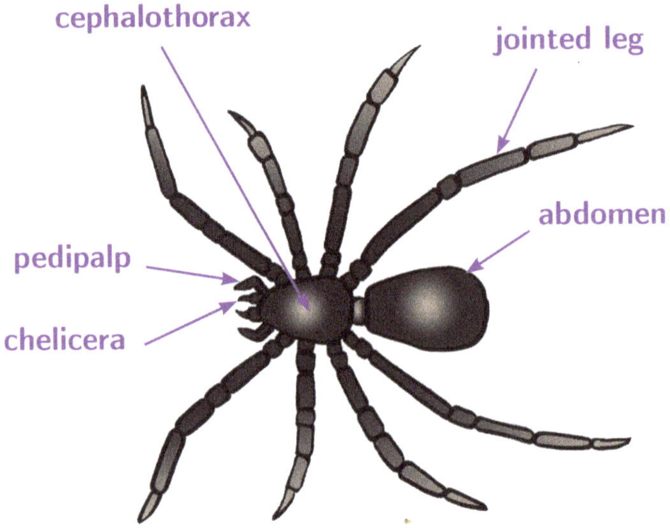

One of the most fascinating features of spiders is their ability to create intricate silk webs. Spiders have **spinnerets** at the end of their abdomen that produce and release silk. The silk inside the spider's abdomen is liquid, but as soon as it contacts the air, the liquid silk turns to a solid. Spiders use silk to create webs to capture prey, make cocoons for their young, and travel from one spot to another. If you are lucky, you may have experienced a spider traveling down from the ceiling to greet you.

8.5 Lobsters, Shrimp, and Crabs

Lobsters, **shrimp**, and **crabs** are in a group of arthropods called **crustaceans**. Although there are fewer types of crustaceans than insects, crustaceans have more varied body parts. Crustaceans have parts for swimming, crawling, burrowing, and jumping.

Lobsters, shrimp, and crabs have five pairs of legs attached to their cephalothorax. One pair of legs often has a set of claws that are used to capture and kill prey.

Most crustaceans live in or near water and can be found in both saltwater and freshwater environments. Unlike insects, the exoskeleton of many crustaceans cannot keep water from escaping, and they can easily dry out if they are away from moisture too long. It's important for most crustaceans to stay in or near water and live in moist environments.

8.6 Summary

- **Arthropods** are the largest group of animals, **Spiders**, **lobsters**, **crabs**, **millipedes**, **flies**, **bees**, **butterflies**, and **ants** are all **arthropods**.
- **Insects** are the largest group of **arthropods**.
- **Insects** have three **segments** — a **head**, **thorax**, and **abdomen**, **six legs**, **compound eyes**, and **antennae**.
- **Spiders** are not insects but are called **arachnids**, and they have a **cephalothorax** (fused head and thorax), an **abdomen**, and **eight legs**.
- **Lobsters**, **crabs**, and **shrimp** have a **cephalothorax** (fused head and thorax), an **abdomen**, and **five pairs of legs**. One pair of legs often contains **claws**.

8.7 Vocabulary

Basic Words

chitin (KIY-tuhn)

glucose (GLOO-kohs)

insect (IHN-sehkt)

thorax (THAWR-aaks)

Challenge Words

abdomen (AAB-duh-muhn)

antenna (aan-TEH-nuh) **antennae** *[plural]* (aan-TEH-nee)

arachnid (uh-RAAK-nuhd)

arthropod (AAR-thruh-pahd)

caterpillar (KAA-tuhr-pih-luhr

cephalothorax (seh-fuh-luh-THAWR-aaks)

chelicera (kih-LIH-suh-ruh) chelicerae *[plural]* (kih-LIH-suh-ree)

chrysalis (KRIH-suh-luhs)

crustacean (kruh-STAY-shuhn)

Lepidoptera (leh-puh-DAHP-tuh-ruh)

mandible (MAAN-duh-buhl)

metamorphosis (meh-tuh-MAWR-fuh-suhs)

mosquito (muh-SKEE-toh)

pedipalp (PEH-duh-paalp)

spinneret (spih-nuh-REHT)

transparent (traans-PEHR-uhnt)

CHAPTER 9

Swimming, Flying, Scaly, Furry ANIMALS

9.1 Introduction

In the last few chapters we looked at animals that don't have a backbone, such as worms, sponges, jellyfish, snails, sea urchins, insects, spiders, and lobsters. In this chapter we will take a look at animals that do have a **backbone**. Recall that a backbone is a series of interconnected bones that run down the back of some animals and that animals with a backbone are called **vertebrates**. Frogs, fish, reptiles, birds, and mammals are vertebrates and are among the animals that have backbones.

9.2 Fish

Fish are a type of vertebrate that live in both freshwater and saltwater environments and can be found in tiny ponds, running streams, and large oceans. There are many different kinds of fish, but in general, fish are divided into three groups — **fish without jaws**, **cartilage fish**, and **bony fish**.

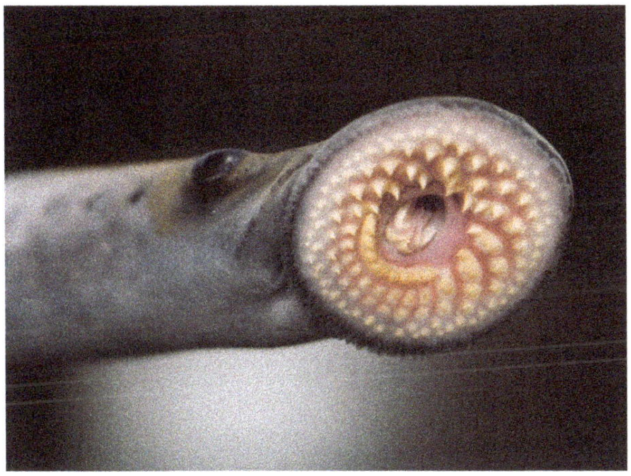

One kind of **jawless fish** is the **lamprey**. Lampreys are found in the open oceans and also in freshwater environments. Lampreys have long thin bodies and resemble eels. Lampreys don't have a jaw but instead consume their food with a rounded sucker-like mouth filled with rows of teeth.

Another type of fish you might see in the ocean is a **shark**. Although sharks might not look like fish, they are a type of fish. Unlike the lamprey, sharks have jaws and often lots of teeth.

Sharks have skeletons made of **cartilage**. Cartilage is a strong but lightweight and flexible tissue. Having a skeleton of lightweight cartilage helps sharks stay afloat instead of sinking. The skin of a shark is covered with sharp, small scales that feel like sandpaper to the touch.

Bony fish are familiar to most people and are the most common type of fish that people catch, eat, and breed for food. **Salmon**, **tuna**, and **trout** are all bony fish. Bony fish are named for their skeletons, which are made primarily of bone rather than cartilage. The scales on bony fish are also made of bone.

How do bony fish stay afloat, swim to the surface, or dive deep? Bony fish have a way to regulate their **buoyancy**. In general, buoyancy is a measure of how well something floats. For example, a rock sinks to the bottom of a glass jar filled with water. Therefore, it doesn't have any buoyancy in water. On the other hand, a piece of wood will float in water, so a piece of wood does have buoyancy in water.

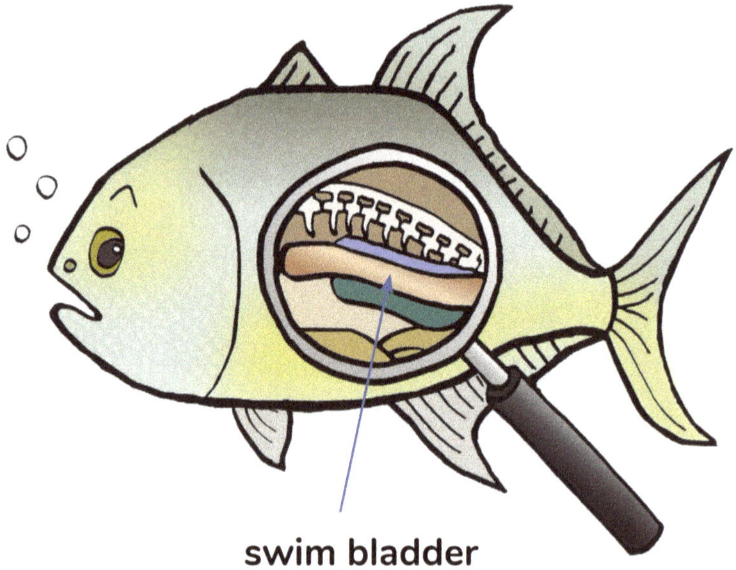

Bony fish have a **swim bladder** inside their abdomen that is like a little balloon. The swim bladder helps them regulate their buoyancy. When they need to go to the surface, they fill the swim bladder with air and float to the top. When they need to dive deeper, they release air so they sink. And when they just want to hang out at a given depth, they can fill the swim bladder with the right amount of air to allow them to swim effortlessly at the depth they choose. Since air is lighter than water, by adding air to their body, the fish become more buoyant and can rise. By letting air out, they become less buoyant and can sink.

9.3 Frogs

Life on Land and in Water

Have you ever thought about what it must be like to live both on land and in the water? Imagine going for a stroll in the park and then having tea with the fish in the pond. It might be fun!

Think about what you might need to have to be able to live on both the land and in the water. What kind of skin would you need? What kind of lungs would you need? What about your feet? What kind of feet would you need?

The word **amphibian** means "both lives." Frogs are called **amphibians** because they live two lives. Most frogs begin their lives as eggs laid in water. The eggs hatch into **tadpoles** that live in the water. As adults, frogs have special skin, lungs, and feet that allow most to live both on land and in water.

The Beginning: Eggs

Most often, a frog begins its life as an egg in water. You probably know that a chicken also starts out as an egg. When you look at a chicken egg you can see that it has a yellow yolk in the middle and a hard outer shell.

A frog egg is much smaller than a chicken egg, but it is not too small to see. If you look closely, you can see that a frog egg also has a yolk. But the yolk of a frog egg is not yellow like a chicken egg. Instead, it is usually black or gray or even black and white.

A frog egg is not covered in a hard outer shell like a chicken egg, but sits in a clump of goo. The goo looks like clear Jell-O. This goo is sticky and helps the eggs stay attached to plants or logs in the water where the mother frog puts them. The goo also helps keep the eggs safe and stuck together so that they don't get lost.

The Middle: Tadpoles

Once the mother frog lays the eggs, she swims away. The eggs hatch on their own. It takes anywhere from a few days to several weeks for the frog eggs to hatch. When they hatch, little **tadpoles** emerge from the eggs in the goo. The tadpoles eat the goo for food until they get a little bigger.

Newly hatched tadpoles are small and fragile. They stick themselves to strands of grass until they get stronger. Tadpoles have gills that help them breathe under water and tails that help them swim. Once they grow a little bigger, they swim around eating algae and plants until they are ready to turn into adult frogs.

The Change: Tadpoles to Frogs

After the tadpole has been swimming and eating and growing, it is ready to become a frog. The change from tadpole to frog is called **metamorphosis,** like the change from caterpillar to butterfly. But unlike a butterfly, a tadpole does not need to be in a chrysalis while it is making the change into a frog. It changes as it swims around, eating and being a tadpole.

The metamorphosis starts when the tadpole begins to grow hind legs. The hind legs start out tiny, then grow bigger and bigger. After the hind legs grow, the front legs appear. The front legs have been growing underneath the skin and, when they are ready, they pop out fully formed. When the

front legs pop out, the tail starts to shrink back into the body. The mouth changes shape and, for some types of frogs, teeth form. The gills disappear and lungs develop so the frog can breathe air and live on land. The tadpole has now become an adult frog.

The End: Adult Frogs

The adult frog looks very different from the tadpole. It can sit and hop and move around on land. It has lungs to breathe air, and it has eyes and even a kind of ear to hear sounds. Most adult frogs have a long, sticky tongue to catch flies and other insects.

There are lots of different kinds of frogs, including Leopard Frogs that have strong legs for jumping and African Clawed Frogs that have claws on their hind feet.

Frogs can be different colors. There are green frogs, brown frogs, blue frogs, red frogs, and yellow frogs. The banana frog is pale green and yellow and has funny looking feet used for climbing.

9.4 Reptiles

Reptiles are a large and diverse group of animals that includes **snakes**, **lizards**, **turtles**, **alligators**, and **crocodiles**. Many reptiles live on land, and a few reptiles, like sea turtles, spend a large amount of time in the water.

Reptiles were once thought to be in the same group as amphibians, but biologists have discovered that reptiles are quite different from amphibians. For example, unlike amphibians, most reptiles lay eggs with thick leathery skins. Because the eggs won't dry out, reptiles don't need to be near water to lay eggs. This allows many reptiles to live in dry climates, like deserts.

Reptiles must use the Sun's energy to stay warm. Reptiles are sometimes called "cold blooded" but this term isn't accurate. Because a reptile's body does not have a way to keep itself at a consistent temperature like our body does, a reptile's blood can be colder than ours. However, if a reptile has been basking in the warm sun for hours, its blood can be much warmer than ours.

Snakes are a common type of reptile found in many different parts of the world. Snakes have elongated bodies, no legs, and a scaly skin that they must shed as they grow and become too large for it. Most species of snakes are found on land, but a few species live in water.

Lizards can appear to be similar to snakes because, like snakes, lizards have scales on their skin. But unlike snakes, lizards have legs they can use to walk, run, jump, and climb up walls. Some lizards, like geckos, have special pads they use to walk upside down.

Alligators and **crocodiles** look like big lizards with large sharp teeth. Both alligators and crocodiles are reptiles that live on land and in water. Alligators and crocodiles are very similar, making it hard to tell them apart. The main differences between the two are the shape of the snout and size of the jaw. Alligators have a U-shaped snout and wider jaws than crocodiles, while a crocodile's snout is V-shaped and the jaws are narrower.

Turtles are another reptile that can be found both on land and in water. Turtles are different from all other reptiles because turtles carry their home with them. Turtles have a hard shell of bony plates that covers their soft body. If a turtle gets frightened, it will pull its head and all four of its legs inside the shell. This can make it difficult for a predator to eat a turtle.

9.5 Birds

Birds can be found in almost every part of the world. Some birds, like penguins, live their lives on the cold Antarctic ice. Other birds, like the colorful toucan, prefer warmer, tropical climates. Many birds can fly, but some birds, like the ostrich, don't fly. Instead, an ostrich uses its long thin legs to run.

With over 9,000 different species of birds, there are so many different types of birds that it is difficult to list them all. However, all birds have some common features. All birds have wings, even if some birds don't use them to fly. All birds have hollow, honeycombed bones. This bone structure helps make birds light enough to fly. All birds are covered with feathers, and all birds have beaks with no teeth.

9.6 Mammals

Did you know that you are a **mammal**? Cats, dogs, mice, whales, and humans are all mammals. But this doesn't mean that all mammals have exactly the same features. A few mammals, like the platypus, lay eggs. Some mammals, like horses, have hooves instead of feet. Humans are the only mammal that walks upright on two legs.

All mammals have two common characteristics. First, all mammals have hair. Some mammals have lots of hair and some have hardly any hair.

Dolphins are born with a few hairs on their chin. As they grow to adults, this hair falls out. Because dolphins have a layer of fat under their skin to keep them warm, they don't really need hair.

The second characteristic of mammals is that all mammals have **mammary glands** with which they feed their young.

There are three main groups of mammals. The first group includes those mammals that lay eggs, like the duck-billed platypus and the Australian spiny anteater.

The second group includes all mammals that use a **pouch** to help nurture their young, like kangaroos, certain species of mice, and opossums.

The third group includes all other vertebrate animals, like cats and dogs, rabbits, sheep, horses, humans, and even bats.

9.7 Summary

- **Fish** are **vertebrates** that live in water. The three groups of fish are **fish without jaws**, **cartilage fish**, and **bony fish**.
- **Frogs** undergo **metamorphosis** when they change from a **tadpole** to an adult **frog**.
- **Frogs** are called **amphibians** because they live part of their lives in water and part on land.
- **Reptiles** include **snakes**, **lizards**, **turtles**, **alligators**, and **crocodiles**.
- **Birds** have wings, feathers, and hollow, honeycombed bones.
- **Mammals** are animals that use **mammary glands** to feed their young.

9.8 Vocabulary

Basic Words

bladder (BLAA-duhr)

dolphin (DAHL-fuhn)

lizard (LIH-zuhrd)

mammal (MAA-muhl)

reptile (REHP-tiyl)

salmon (SAA-muhn)

tadpole (TAAD-pohl)

tuna (TOO-nuh)

turtle (TUHR-tuhl)

Challenge Words

alligator (AA-luh-gay-tuhr)

amphibian (aam-FIH-bee-uhn)

buoyancy (BOI-uhn-see)

cartilage (KAHR-tuh-lihj)

crocodile (KRAH-kuh-diy-uhl)

mammary gland (MAA-muh-ree GLAAND)

metamorphosis (meh-tuh-MAWR-fuh-sihs)

vertebrate (VEHR-tuh-bruht)

CHAPTER 10

What Are WAVES?

10.1 Introduction

The next time you take a bath, notice what the water looks like. With the tub full and the water faucet turned off, notice that the water is calm and smooth. Then put your foot in the tub and watch what the water does. If you observe carefully, you can probably see **waves** of water moving away from your foot toward the sides of the tub and maybe even bouncing back.

Notice that if you put your foot into the tub slowly, you make small waves, and if you put your foot into the tub quickly, you make bigger waves. You can make small waves or big waves depending on how fast you make the water move.

Also notice that the waves change over time. If you put your foot in the water quickly, the wave might be big and move fast. But if you keep watching the water, you can see that the waves slow down and eventually disappear.

Waves can be big or small, fast or slow, and they can change over time.

10.2 Parts of a Wave

If you look carefully at a wave, you can see that it has a high part and a low part. Scientists call the high part of the wave the **peak**, and the low part of the wave the **valley**.

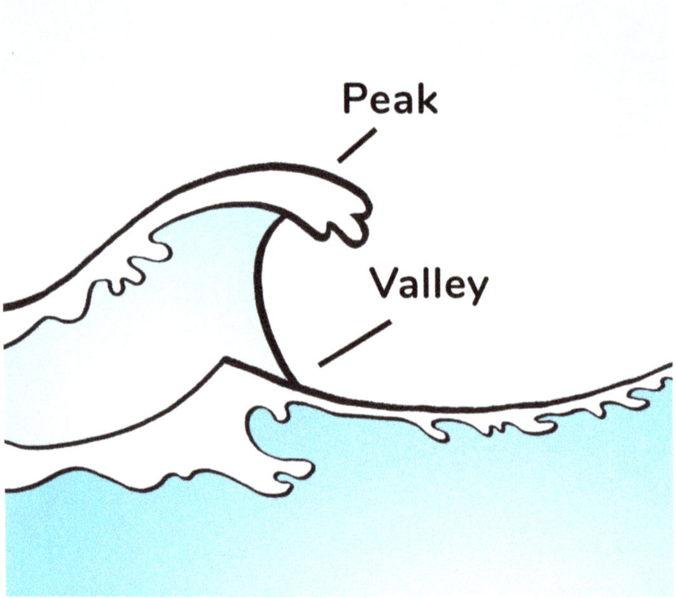

If we draw the wave on a piece of paper, it might look something like this:

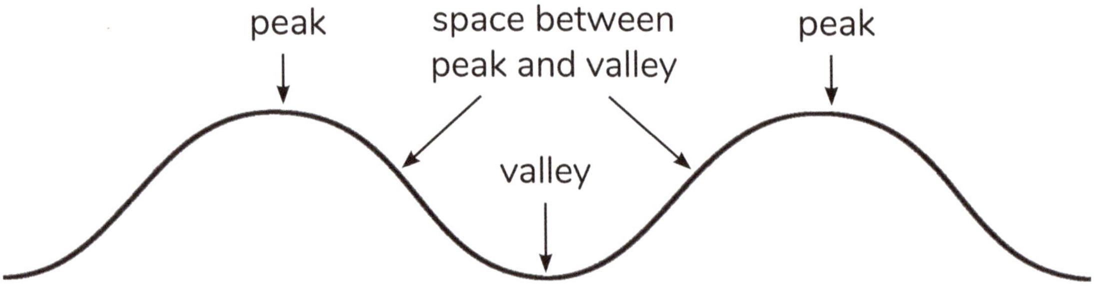

Taking a closer look at the wave, we can see that the peaks and valleys have space between them.

Cutting the wave in half lengthwise (shown by the dotted line), we can label some of these parts. The height of the wave measured from the middle is called the **amplitude**, the distance between the peaks is called the **wavelength**, and how fast the waves move is called the **frequency**. We will take a closer look at frequency in the next section.

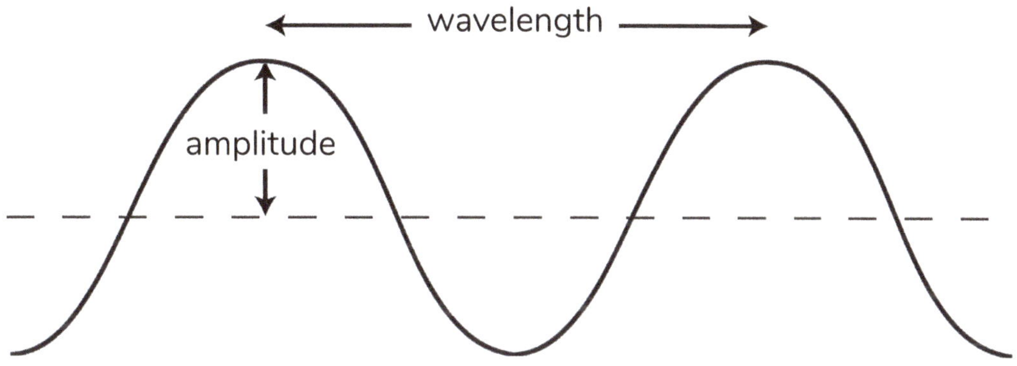

If we stretch out the wave horizontally, the distance between the peaks grows and the wavelength gets longer, but the amplitude stays the same.

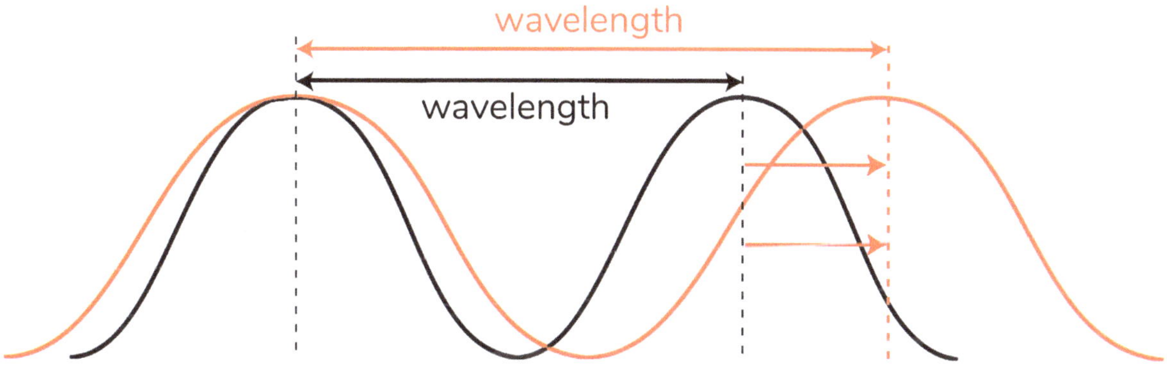

If we squeeze the wave horizontally, the distance between the peaks decreases and the wavelength gets shorter, but the amplitude stays the same.

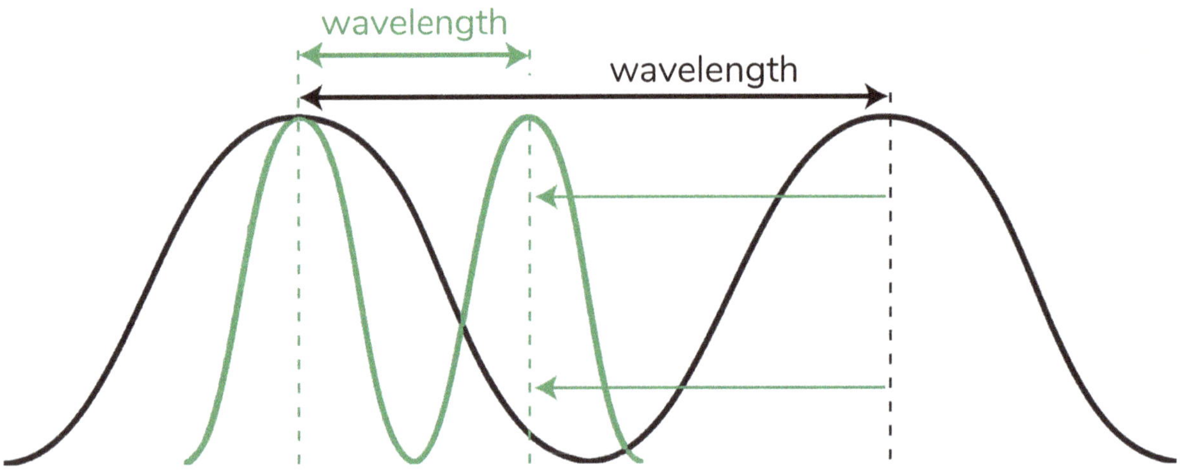

If we stretch the wave vertically, the height of the peaks grows and the amplitude increases, but the wavelength stays the same.

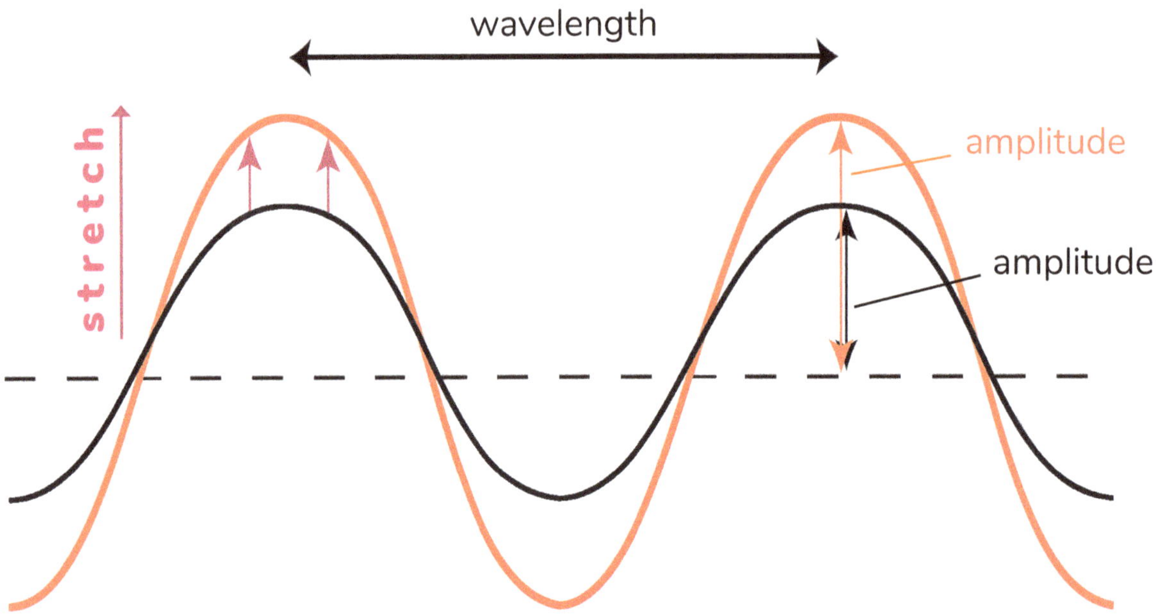

If we squeeze the wave vertically, the height of the peaks gets smaller and the amplitude decreases, but the wavelength stays the same.

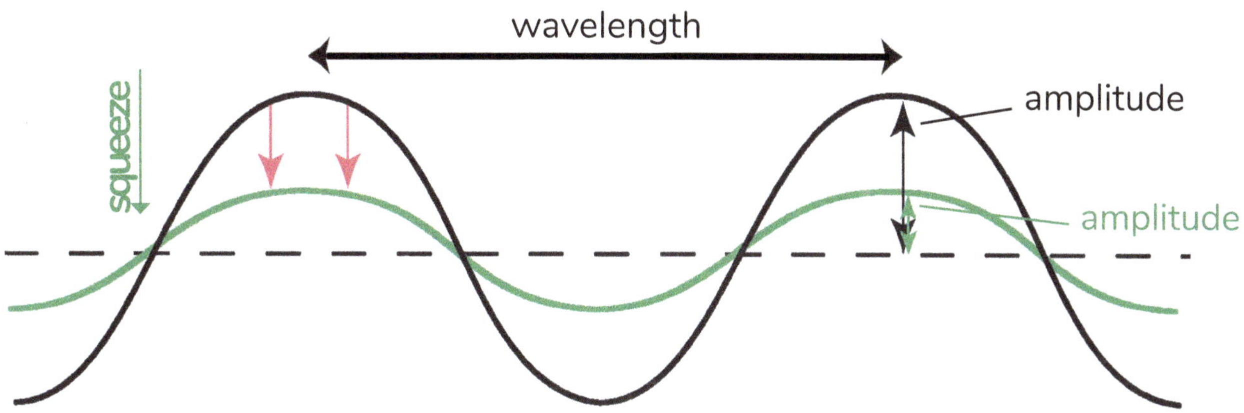

By stretching and squeezing a wave, the wavelength and the amplitude can be increased or decreased. Understanding how waves change is important in determining how waves behave, how fast they can travel, and where they can go.

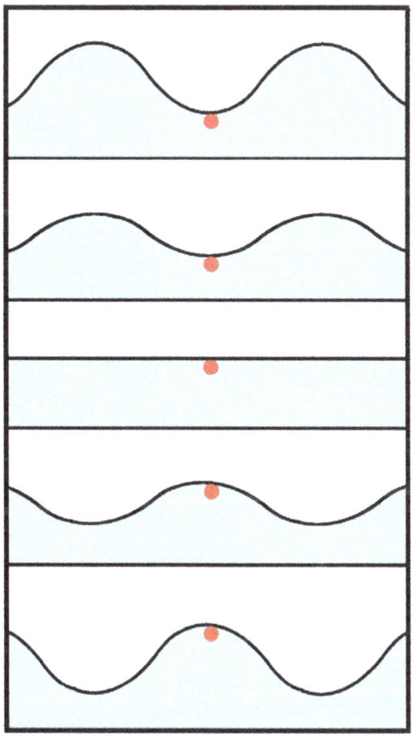

10.3 Waves Oscillate

Another important property of waves is that they **oscillate**, or move back and forth. If you are watching a water wave and you keep looking at the same spot, you will see the water go up and down. This is oscillation. A guitar string also moves in a similar way, but it oscillates much faster.

The speed of oscillation is measured by counting how many times the wave bounces up and down in one second. This is called the **frequency of oscillation**. One complete bounce of a wave is called a **cycle**. A water wave may bounce up and down twice in a second, so it has a **frequency** (frequency of oscillation) of 2 cycles per second. A plucked guitar string may bounce about 100 times per second, so its frequency is about 100 cycles per second. The guitar string has a higher frequency than the water wave.

In summary, every wave has three properties:

1. **Amplitude** — how *high* is it?
2. **Wavelength** — how *big* is it (from peak to peak)?
3. **Frequency** — how *fast* is it?

- **In a water wave**, the amplitude of a wave determines how high the wave is. The wavelength determines how far apart the peaks of the waves are, and the frequency determines how fast the waves are traveling.

- **In a sound wave**, the amplitude of the wave determines how *loud* the sound is, and the wavelength and frequency together determine the *pitch* of the sound. A high-pitched sound has a short wavelength and a high frequency. A low-pitched sound has a long wavelength and a low frequency. We will learn more about sound waves in Chapter 11.

- **In a light wave**, the amplitude of the wave determines how *bright* the light is. The wavelength and frequency together determine the color of the light wave. We will learn more about light waves in Chapter 12.

10.4 Waves Transfer Energy Not Matter

Going back to the example of putting your foot in the bathtub, when the wave moves across the tub, it looks like the water is moving forward. But if you could look at the individual water molecules, you would see that they aren't actually moving with the wave. They are moving in one place, either back and forth or up and down.

A good way to think about how waves work is to observe a **stadium wave**. If you've ever been to a football or basketball game, you might have observed a moment where suddenly a section of the audience stands up and then sits down again. After they sit down, the section right next to them stands and sits down, and the next section does the same, all around the stadium. This is called a stadium wave and it's lots of fun!

If you are observing the audience from across the field, and if the people in the audience are taking turns standing and sitting at the correct times, you can observe how the people are moving in a wave. You can observe how the individual audience members are temporarily **displaced** from their resting position, but they don't actually change horizontal locations (they don't walk around the stadium and move to a different seat). They are just standing or sitting in the same place.

In the same way, a wave temporarily displaces molecules as it **transfers energy** from one place to another. Waves can transfer energy through water, air, wood, and other materials. A wave doesn't actually carry the molecules in the materials along with it. The wave just temporarily displaces the molecules. The molecules are temporarily disturbed as the wave transfers energy from one place to another.

10.5 Types of Waves

There are many different ways to categorize waves. One way is to look at how the molecules in the wave are displaced.

A **transverse wave** is a wave in which the molecules in a material move **perpendicular to the direction the wave is traveling**.

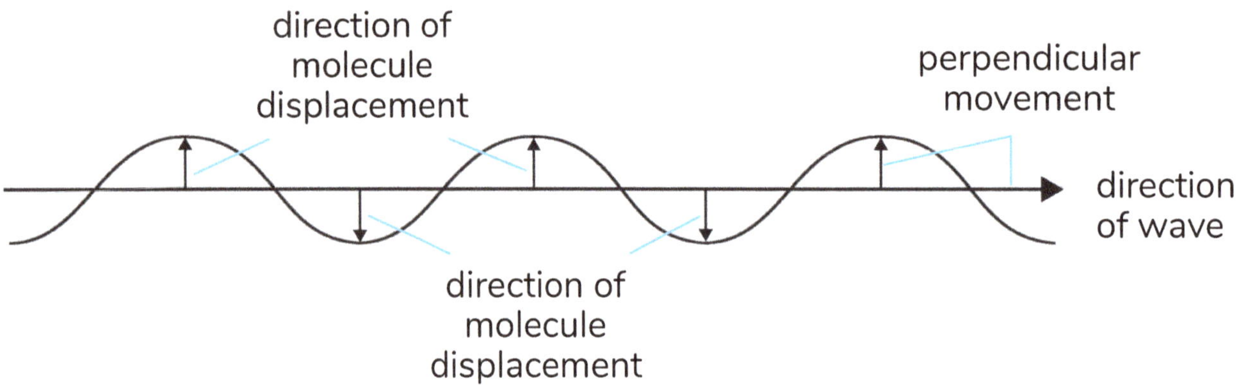

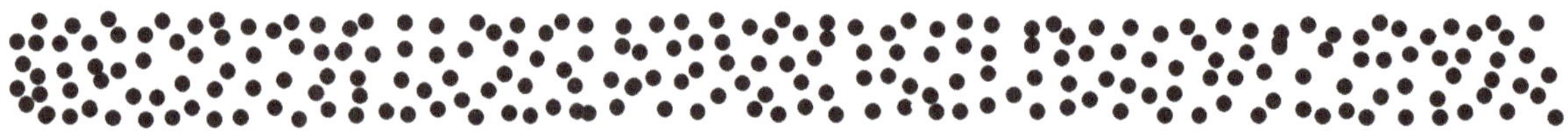

Molecules before a wave passes through them

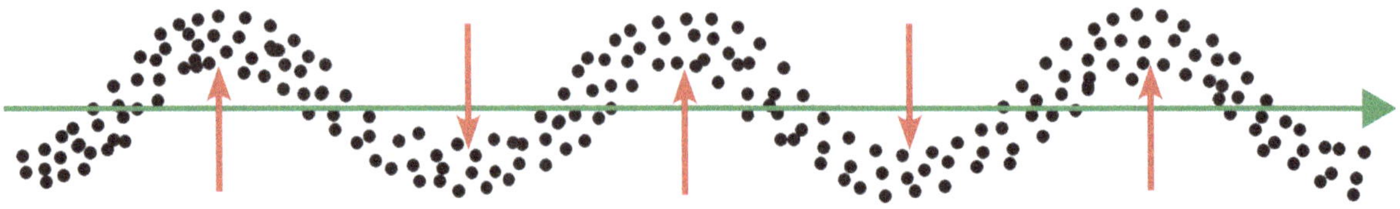

Molecules are displaced as a wave passes through them

Imagine holding one end of a rope and having a friend hold the other end tightly. If you start to wiggle your end of the rope, you will create a transverse wave. You will notice that the wave moves down the rope from one end to the other. Your wiggling of the rope creates the starting point for the wave. This starting point is where energy is being transferred from your hand to the rope. The ending point is were the wave runs out of rope to displace. Your friend feels the tugging of the rope from the energy being transferred. The rope fibers (the molecules in the rope) move **perpendicular** to the rope, but the rope fibers on one end don't move along with the wave to the other end of the rope.

Another way to say this is that the rope fibers, or molecules, are being displaced and are not moved along the rope from one end to the other.

A **longitudinal wave** is a wave where the molecules are **displaced parallel to the direction the wave is moving**. The molecules are not carried along with the wave but only move slightly back and forth as the wave passes through them.

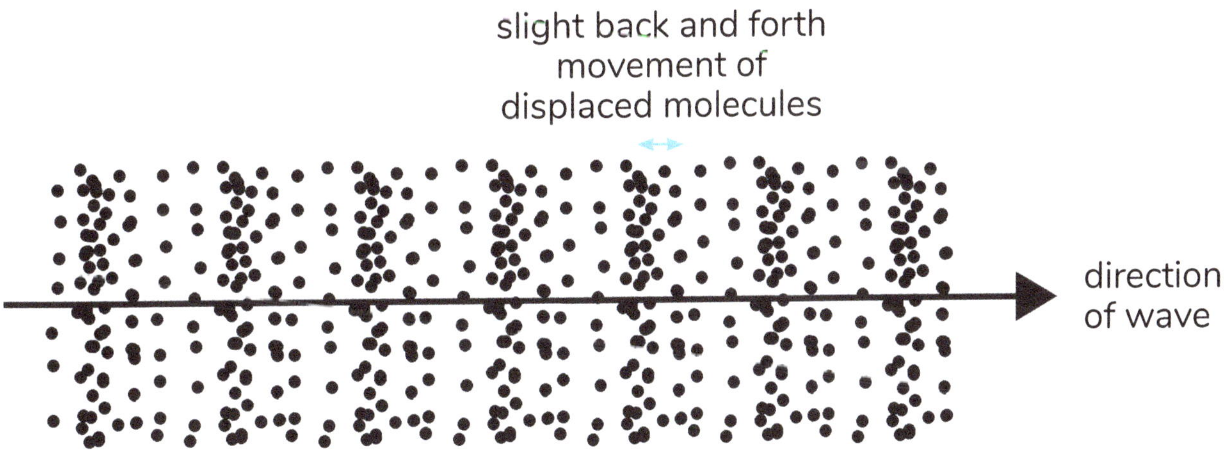

Imagine having two friends stretch a Slinky between them. Imagine a third friend placing their fingers between the coils and squeezing the coils together. When the third friend lets go of the coils, you can watch as the Slinky moves back and forth. This is a longitudinal wave because the material in the coil oscillates in the same direction that the wave is traveling.

- A **circular wave** is a wave where the molecules **move outward from the center** in a circular motion.

The waves on the surface of the ocean are circular waves. Waves deep down in the ocean are longitudinal waves. If you dive below the surface of the ocean, you can feel the water push you gently back and forth. This is the longitudinal wave of the ocean. However, if you swim to the surface and a big wave comes up, you might get tossed about by the wave, and it might just roll you over! This is because surface waves in the ocean move the water in a circular motion.

- Waves moving through the ocean, through a rope, or through the air are called **mechanical waves**. Mechanical waves need some type of matter to move through, like water, rope fibers, or air molecules. Mechanical waves are nothing more than a disturbance in matter.

Waves in a bathtub are a disturbance in the water, and sound waves are a disturbance in the air. Without matter, mechanical waves can't transport energy.

However, there is a type of wave that can transport energy without matter. **Electromagnetic waves** can transport energy through a **vacuum** (empty space). An electromagnetic wave is a special combination of **electric and magnetic fields**.

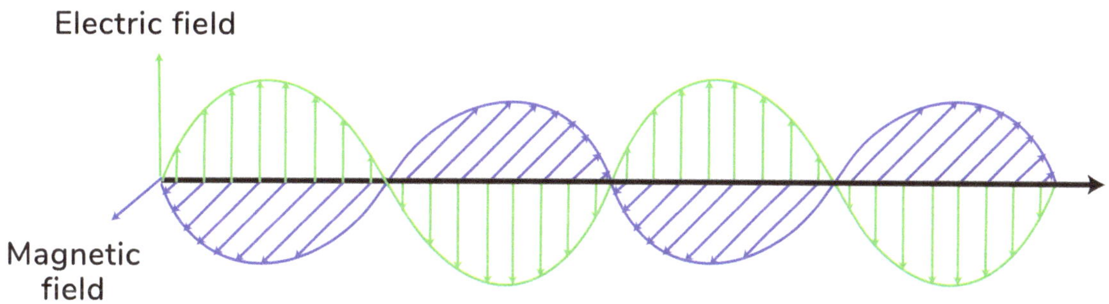

Electromagnetic wave

Electromagnetic waves are made when charged particles, like electrons, vibrate and create electric and magnetic fields. Light waves, radio waves, UV, and infrared waves are all electromagnetic waves. The only difference between these waves is their wavelength. Radio waves, microwaves, and infrared light have longer wavelengths. Ultraviolet, x-rays, and gamma rays have shorter wavelengths. **Visible light** (light that we see with our eyes) has wavelengths shorter than radio waves and longer than x-rays. We will learn more about visible light in Chapter 12.

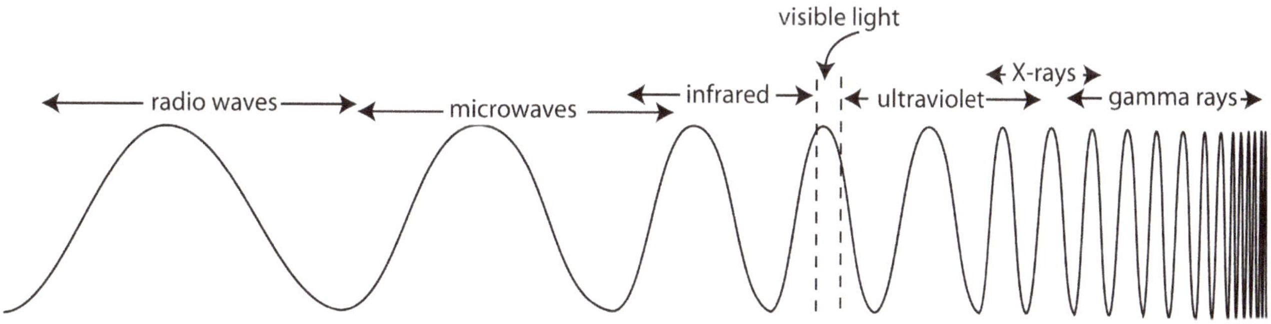

Electromagnetic spectrum

10.6 Summary

- **Waves** have **peaks** and **valleys**.
- The distance between the **peaks** of a **wave** is called the **wavelength**.
- The height of a **peak** is called the **amplitude**.
- **Waves** transfer energy, not matter.
- A **transverse wave** is one where the molecules move **perpendicular to the direction the wave is traveling**. The molecules are not carried along with the wave.
- A **longitudinal wave** is one in which the molecules move **parallel to the direction the wave is moving**. The molecules are not carried along with the wave but only move slightly back and forth as the wave passes through them.
- **Surface or circular waves** transfer energy through matter by moving molecules in a **circular motion**. The molecules are not carried along with the wave.
- **Mechanical waves** need matter to carry them.
- **Electromagnetic waves** can transfer energy in a **vacuum** where there is no matter.

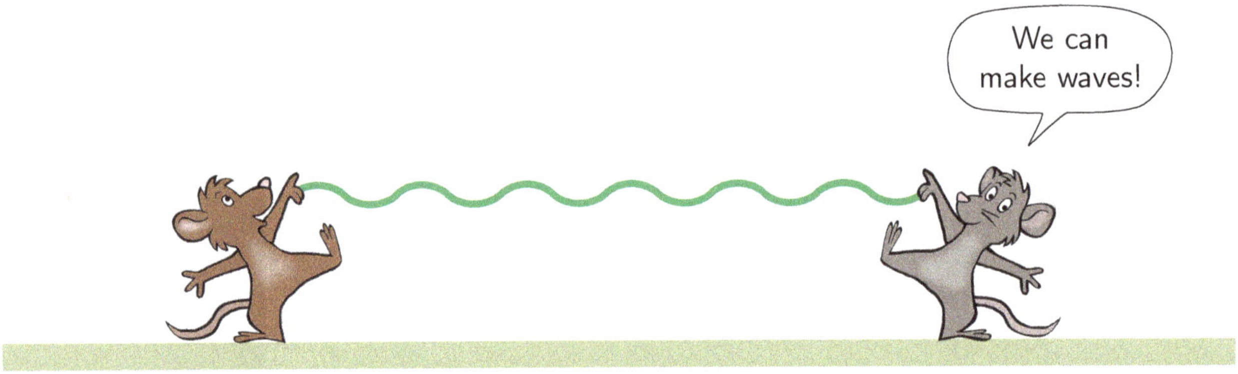

10.7 Vocabulary

Basic Words

displace (dihs-PLAYS)

wave (WAYV)

Challenge Words

amplitude (AAM-pluh-tood)

circular (SUHR-kyuh-luhr)

electric (ih-LEHK-trihk)

electromagnetic (ih-lehk-troh-maag-NEH-tihk)

frequency (FREE-kwuhn-see)

longitudinal (lahn-juh-TOOD-nuhl)

magnetic (maag-NEH-tihk)

mechanical (mih-KAA-nih-kuhl)

oscillate (AH-suh-layt)

parallel (PEHR-uh-lehl)

perpendicular (puhr-pehn-DIH-kyuh-luhr)

stadium (STAY-dee-uhm)

transverse (traans-VUHRS)

visible (VIH-zuh-buhl)

wavelength (WAYV-lehnkth)

CHAPTER 11

What Is SOUND?

11.1 Introduction

If you clap your hands what happens? You hear a sharp sound and then it fades. Clap them again. Again, you hear a sharp sound that soon fades.

When you clap your hands, you are suddenly moving the air molecules near your hands and creating a wave. The wave continues to move through the air molecules and reaches your ears. Your ear drums pick up the vibration of the moving air molecules, and you hear a sound. When you are finished clapping, the air molecules that were bouncing around settle, and the sound fades.

- **Sound** is a mechanical wave that moves **longitudinally** through matter. Sound will travel through air, water, wood, steel, and guitar strings. Physicists refer to the material a sound wave travels through as the **medium**. Air is a medium. Water is a medium. Both wood and steel are mediums.

11.2 Different Kinds of Sound

What is the difference between the sound of a bird chirping and the sound of a bass drum? Both are sound waves traveling through the air. But a bird chirping sounds different from a bass drum. A bird chirping sounds sharp and high, and a bass drum sounds dull and low.

The difference between a bass drum and a bird chirping is the **pitch** of the sound. Recall that sound waves oscillate. Pitch is related to the **frequency** of the sound wave. A high pitched sound has a high frequency, and a low pitched sound has a low frequency.

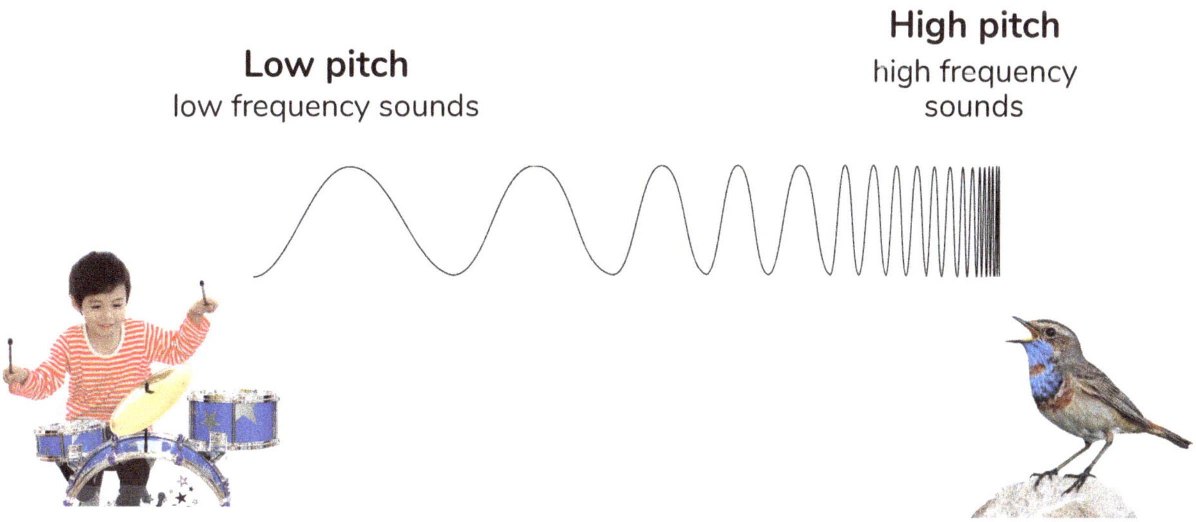

Looking at the wave itself, frequency is the number of peaks that occur over a given period of time. High frequency sounds have shorter wavelengths, and low frequency sounds have longer wavelengths.

Frequency is related to how quickly the medium vibrates as the wave passes through it. For a high frequency sound, the medium is vibrating quickly, and for a low frequency sound, the medium is vibrating slowly. The frequency of a sound wave is exactly the number of times per second that the particles in a medium move back and forth.

The mathematical formula for calculating frequency is:

frequency = number of vibrations ÷ time

- For example, if the molecules in air move back and forth 100 times in 5 seconds, then the frequency of the sound would be 100 vibrations divided by 5 seconds, which equals 20 vibrations per second.

  frequency = 100 vibrations ÷ 5 seconds = 20 vibrations per second

- If the molecules in air move back and forth twice as fast (200 times in 5 seconds), the frequency would be higher (40 vibrations per second).

  frequency = 200 vibrations ÷ 5 seconds = 40 vibrations per second

- If the molecules in the air move back and forth half as fast (50 times in 5 seconds), the frequency would be lower (10 vibrations per second).

  frequency = 50 vibrations ÷ 5 seconds = 10 vibrations per second

Human ears can detect small differences in frequency. In fact, highly trained musicians can detect differences as small as 2 vibrations per second.

Sounds can also be loud or soft. We often think of this as *volume*, but in physics this is called **intensity**. The intensity of a sound depends on the amplitude of the sound wave. The higher the amplitude, the louder the sound. The lower the amplitude, the softer the sound.

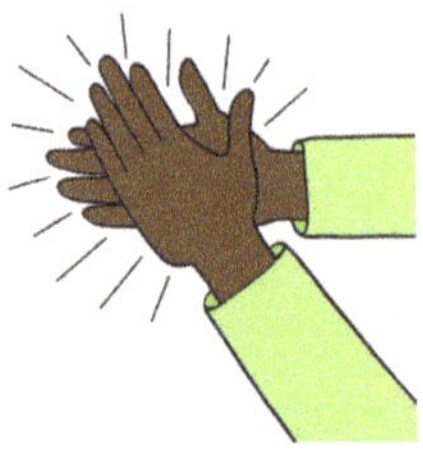

Soft clap
low intensity

Hard clap
high intensity

The intensity of sounds is measured in a unit called a **bel**, named after **Alexander Graham Bell**. Bell was born in 1847 in Edinburgh, Scotland and later moved to the United States. Bell invented the telephone by using his understanding of sound and electronics. In his first successful experiment, he startled his young assistant, Mr. Watson, when from another room, he asked Mr. Watson to "come here."

Sound intensities are usually measured in **decibels**. Ten times the intensity of 0 bels is one bel or 10 decibels, and 100 times more intensity than 0 bels is 20 decibels. The minimum level of sound that a human ear can hear, which is called the **threshold of hearing**, is 0 decibels. A whisper is about 20 decibels, and rock music can be about 110 decibels. Hearing can actually be damaged by sounds above about 85 decibels, depending on the length of exposure and frequency of the sound. This is why people working around loud machinery or at airports wear protection on their ears.

11.3 How Fast Does Sound Travel?

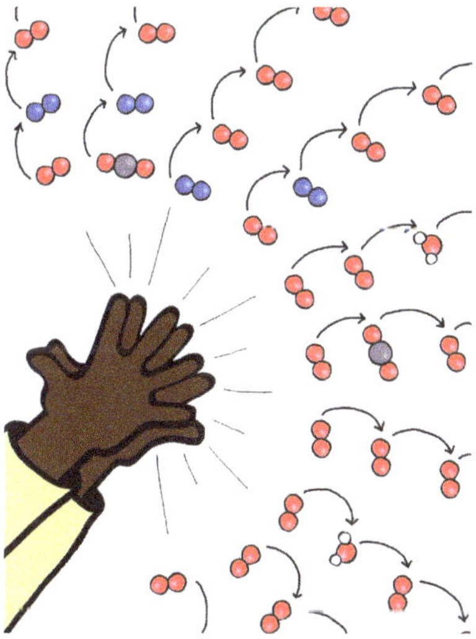

We know that sound is a mechanical wave that travels through space longitudinally. But what does this mean, exactly?

Recall that when you clap your hands, you are disturbing the air molecules surrounding your hands. These disturbed molecules wiggle around for a while and then settle down. But before they settle down, they pass the energy of their movement to the molecules nearby. These new molecules wiggle for a while and then pass their energy to the molecules near them before settling down.

Recall what happens when you drop a pebble in water. The pebble disturbs the water nearby, creating a wave that moves outward from the pebble. In the same way, your hands disturb the air molecules surrounding your hands, creating a wave that moves outward from your hands.

How fast does the sound wave get to your ear when you clap your hands? It depends on the type of molecules the sound wave is traveling through. In general, sounds travel faster through solids, less fast through liquids, and least fast through gases.

Why is this? Recall that the molecules in a sound wave interact with surrounding molecules. In a solid, the molecules are more closely packed than in a liquid or a gas. When molecules are more closely packed, it is easier for the molecules to transfer their energy to other molecules.

In order for sound to travel quickly through a solid, the solid also has to be **rigid** with organized atoms. A soft solid won't work as well. For example, a rigid material like steel isn't easy to deform or bend. You can't bend steel with your bare hands, unless you have superpowers. When a sound wave travels through a rigid material such as steel, the disturbance can travel very quickly. Because the molecules are closely packed in steel, the energy moves from one atom to the next very quickly. However, solid materials that aren't rigid, such as certain plastics, rubber, or Styrofoam, do not allow sound to travel quickly. This is because their molecules are not as closely packed and are somewhat disorganized.

Speed of Sound per Second		
Material	**Meters**	**Feet**
Rubber	60	197
Air	343	1125
Water	1497	4911
Aluminum	6320	20,735

In rubber, sound travels about 60 meters (197 feet) per second. In air, sound travels at 343 meters (1125 feet) per second. In water, sound travels at 1497 meters (4911 feet) per second, and in aluminum, sound travels at 6320 meters (20,735 feet) per second.

11.4 What Happens When Sound Bounces?

If you drop a big rock into the middle of a pond, you can see water waves traveling outward from where the rock entered the water. If nothing is in the way, the waves will travel outward until they eventually disappear. But what happens to water waves if they hit a stone wall or some other obstacle in their path?

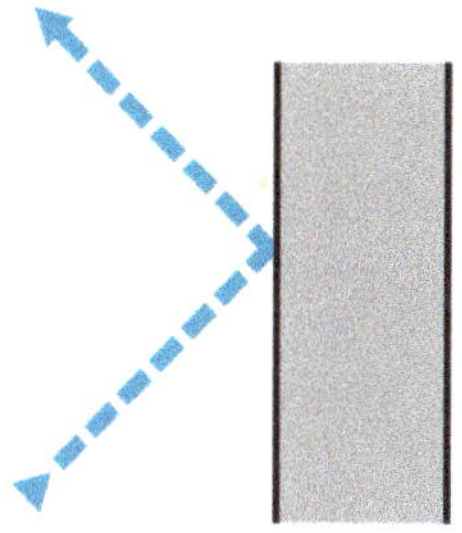
reflection

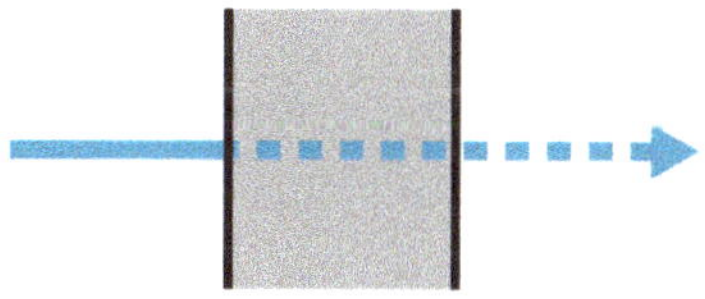
transmission

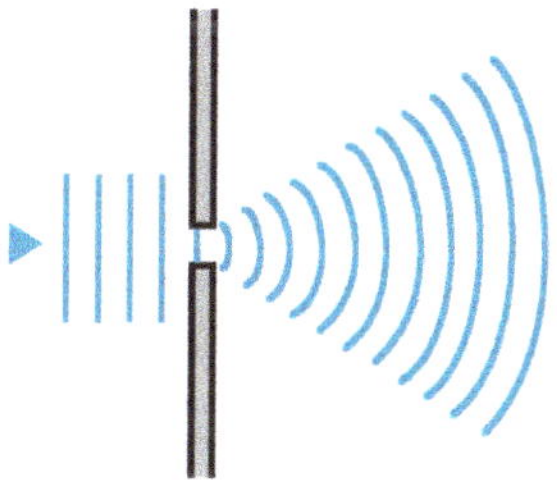
diffraction

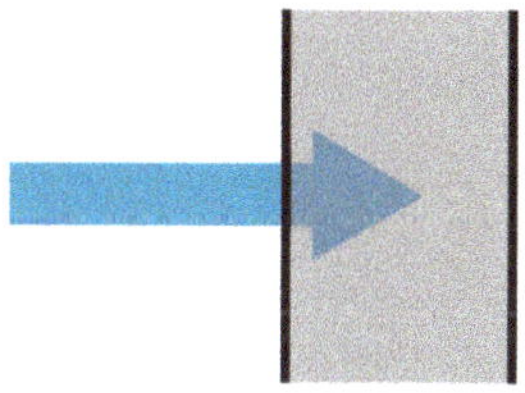
absorption

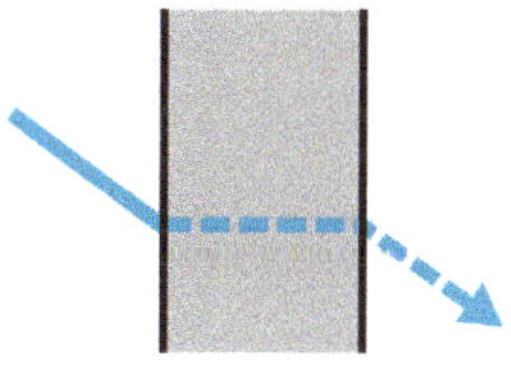
refraction

Waves, including sound waves, will interact with an obstacle in one of several ways. Sometimes a wave will **reflect**, or bounce back, away from an obstacle it hits. At other times a wave will be **transmitted**, or moved, through the obstacle. Sometimes waves can be **diffracted**, which means they can change their direction, travel around corners, travel around obstacles, or move through openings. Sometimes waves are **absorbed**, or taken in, by the material and turned into heat, and sometimes waves are **refracted**, causing the path of the wave to bend and change speed.

When a sound wave encounters an obstacle, whether it is reflected, transmitted, diffracted, absorbed, or refracted is determined both by the material it is traveling through and the material it encounters.

If you walk through a canyon, tunnel, or large room, you might notice that your voice echoes. An **echo** happens when sound waves are reflected, or bounced back, from the walls of the canyon, tunnel, or room. When you speak, your voice travels through the air and hits the walls. Then the sound bounces back and you hear your voice again.

Soft materials can be used to absorb sound and prevent echoes. Rooms that are used for recording music often have thick acoustic tiles made of fiberglass that can absorb sound and keep it from reflecting back into the microphones during a song.

When echoes (reflected sound waves) occur in poorly designed concert halls, they can make it miserable for both the band and the audience.

However, not all reflected sound waves make concerts a miserable experience. Sound can be made to reflect in controlled ways. By understanding how sound waves travel, engineers can use different materials to design auditoriums that enrich the audience's experience of the sound.

The study of sound is called **acoustics**. Acoustic engineers use varied materials, rough surfaces, and other devices in concert halls to help the audience experience the full sound of the music being played rather than just hearing echoes.

11.5 Summary

- **Sound** is a mechanical wave that moves **longitudinally** through matter.
- **Sound** travels through many different materials, which are called **mediums**.
- **Pitch** is related to the **frequency** of the **sound wave**. A high pitched **sound** has a high frequency, and a low pitched **sound** has a low frequency.
- **Sound** travels at different speeds in different **mediums**.
- **Sound** can be **reflected**, **transmitted**, **diffracted**, **absorbed**, and **refracted**.

11.6 Vocabulary

Basic Words

absorb (uhb-SAWRB)

diffract (dih-FRAAKT)

echo (EH-koh)

reflect (rih-FLEHKT)

refract (rih-FRAAKT)

rigid (RIH-juhd)

transmit (traans-MIHT)

Challenge Words

decibel (DEH-suh-behl)

frequency (FREE-kwuhn-see)

intensity (ihn-TEHN-suh-tee)

longitudinally (lahn-juh-TOOD-nuh-lee)

medium (MEE-dee-uhm)

threshold (THRESH-hohld)

vibration (viy-BRAY-shuhn)

CHAPTER 12

What Is LIGHT?

12.1 Introduction

Everything we see, we see because of **light**. Our eyes pick up light that interacts with and bounces off objects. We see colors, shapes, sizes, and textures because of light. Without light we can't see anything. But what is light?

Is light a **wave** or a **particle?** This question has been the source of many heated arguments among scientists. Light reflects like a wave. Light can be diffracted like a wave. Light can be absorbed by certain materials in the same way as a wave. We can see that in many ways light behaves like a wave.

But light can also behave like a particle. We can think of light as behaving like tiny bullets flying through space. Like bullets, light particles can be counted and light particles travel through space in a straight line unless a force acts on them. Therefore, experiments done with light show that light also behaves like a particle.

The fact that light behaves like both a wave and a particle can be easily observed by taking a conventional film photograph. When you snap a picture with a film camera, an image is produced when light passes through the lens system and hits the photographic film. The light then interacts with the chemicals on the film to produce an image. As the light passes through the lens of a camera to the film, the light is behaving like a wave.

But when the light reaches the chemicals on the photographic film, it behaves like a particle. As each particle of light hits a point on the photographic film, energy is being absorbed by a single grain (a silver halide crystal) on the film. As more light particles hit the film, more energy is absorbed by more grains, causing the image to be captured on the film.

For film photography, the number of light particles hitting the film determines how well the image comes out. If too little light hits the film, there will too few grains that have absorbed energy from the light and the

Too little light

Correct amount of light

Too much light

image will look spotty, or grainy. If too much light hits the film, too many grains will absorb energy from the light, and the image will be overexposed, looking too light, or washed-out.

12.2 Photons

Light is made of **photons**. A photon is a packet of energy that has a particular **frequency** and travels through space like a wave. Although physicists currently believe that photons actually have no mass, when photons act like particles, they behave as if they do have mass.

The theory of photons has been difficult for physicists to figure out. While trying to understand what photons are and how they behave, physicists developed two new areas of study called **quantum mechanics** and **special relativity**. According to quantum mechanics, photons are the fundamental unit of light in the same way that atoms are the fundamental unit of matter.

Light, which is made of photons, is a type of **electromagnetic radiation**. The word *radiation* used in this way does *not* mean radioactive. Radiation, in this case, just means that energy is given out, or **radiated**, from its source by **electromagnetic waves**. Normal light from the Sun or from a light bulb is radiation in this broad sense.

As we saw in Chapter 10, electromagnetic radiation is made of both electric and magnetic energy which combine to form an electromagnetic wave. In a vacuum all electromagnetic waves travel at the same speed — the **speed of light**, which is 186,000 miles per *second!*

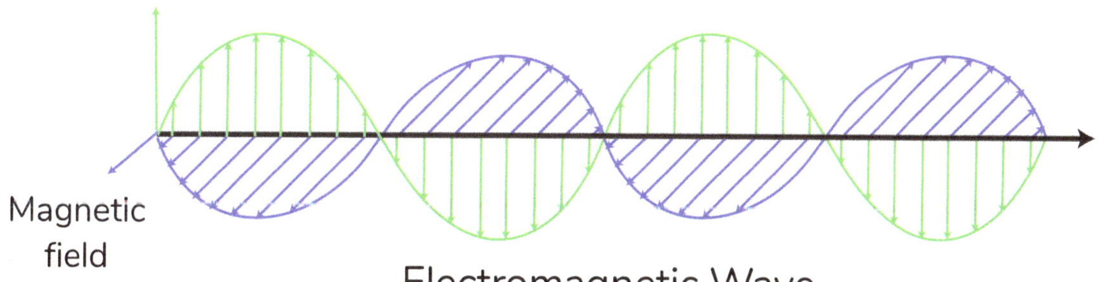

Electromagnetic Wave

The only difference between different forms of electromagnetic radiation is the frequency or **wavelength** of their waves. Radio waves have low frequencies and long wavelengths, and gamma rays have high frequencies and short wavelengths. Visible light has frequencies between radio frequencies and gamma rays.

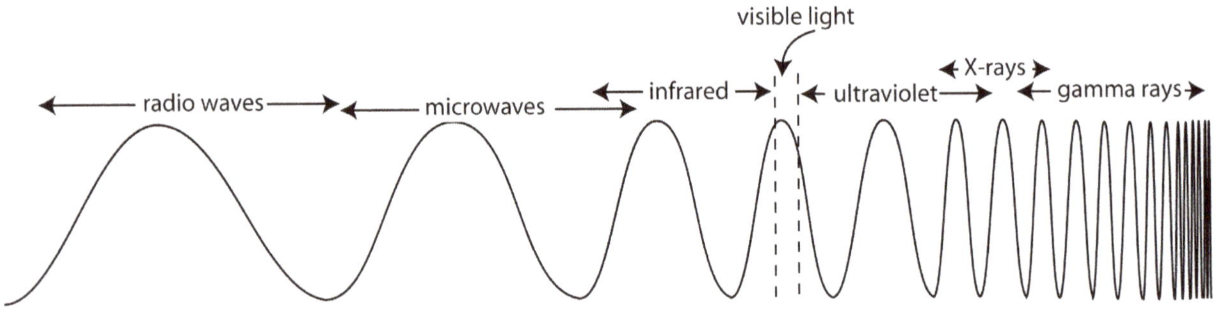

Electromagnetic spectrum

12.3 Visible Light and Color

We cannot see ultraviolet light, and without special equipment we cannot see infrared light, like foxes do. However, the unaided human eye can see a certain range of wavelengths that are in the **electromagnetic spectrum**. These wavelengths are called **visible light**. If we look carefully at the visible part of the spectrum, we find that visible light is made of many different waves at different wavelengths.

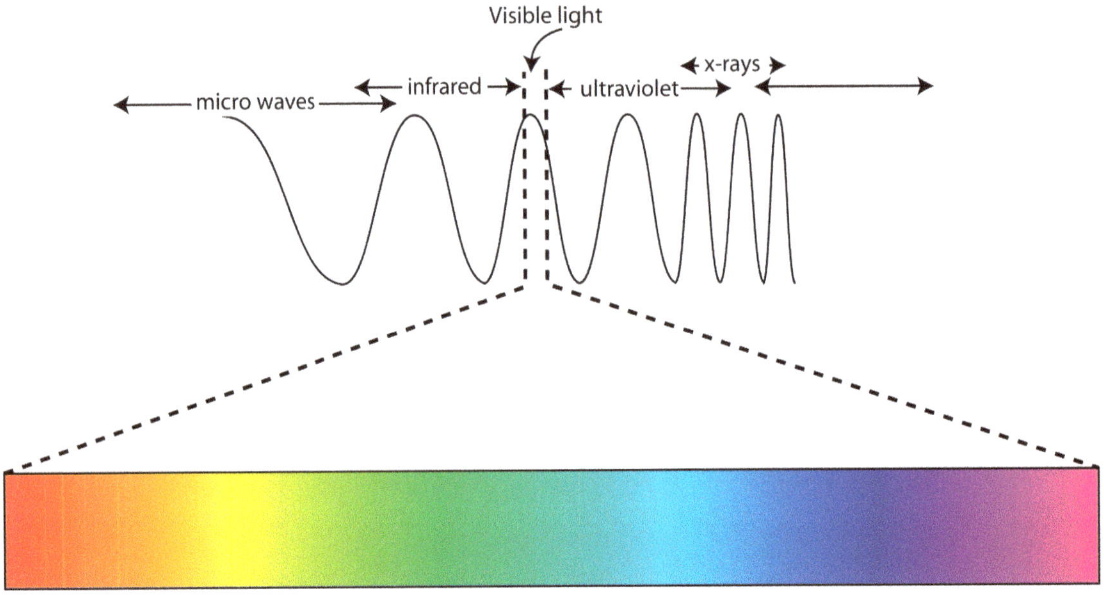

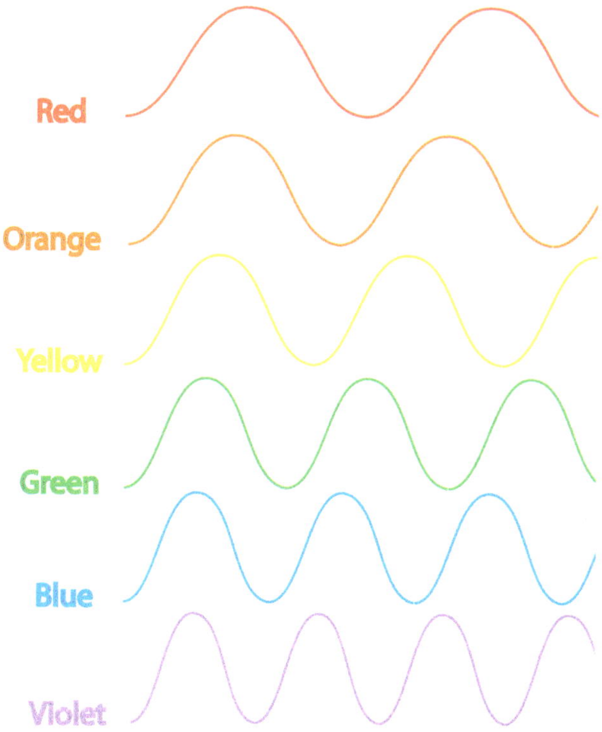

The different wavelengths are the different colors we see. Red light has the longest wavelength in the visible spectrum, and violet light has the shortest wavelength. Orange, yellow, green, and blue are in between red and violet.

It is important to realize that the colors merge continuously into each other. For example, by taking the wavelength of red and squeezing it smaller, it turns into orange. If it is squeezed a little further, it turns into yellow, then green, and so on. This is why the colors look like they run into each other in a rainbow or in the refraction of sunlight through a prism. The colors *do* run into each other. The only difference between them is how long or how short their wavelengths are.

12.4 The "Colors" of Atoms

What makes fireworks so colorful? Why do some fireworks explode with brilliant white sparkles and others explode with red, blue, or purple bursts of light? How do fireworks work?

Fireworks release sound, light, and heat. When a firework explodes, a very fast chemical reaction releases a large amount of energy very quickly. This quick releasing of energy creates a **sonic boom** which results from air molecules moving faster than the speed of sound.

The colors in fireworks are created when this heat energy is absorbed by the atoms of the metallic chemicals in the firework. When metal atoms are quickly heated, the electrons absorb the heat energy and release that

energy as photons. Both metals and metal salts, such as sodium chloride, will **emit**, or release, photons if quickly heated. Different metals emit photons at different wavelengths, which we see as different colors.

Lithium and strontium salts give off photons at long wavelengths that appear red. Sodium salts, such as sodium chloride, give off photons at a slightly shorter wavelength, and they appear as yellow. Barium salts give off photons at a shorter wavelength than sodium, and they appear as green. Copper salts give off photons at an even shorter wavelength. These appear as blue.

COLOR	COMPOUND	WAVELENGTH (in nanometers)
Red	strontium salts and lithium salts	600-646
Orange	calcium salts	591-603
Yellow	sodium salts (sodium compounds)	589
Green	barium salts	511-533
Blue	copper salts	460-530
Purple	mixture of strontium salts and copper salts or potassium salts	432-456

Reference: Imperial College of Science, Tech & Medicine, London, fireworks page; http://www.ch.ic.ac.uk/local/projects/gondhia/lightcolour.html

Metals and metal salts can be mixed to create a variety of colors in a firework. For example, a mixture of strontium and copper will result in a firework that has both red and blue colors.

12.5 Summary

- **Light** is both a **wave** and a **particle**.
- Light is made of **photons**. A **photon** is a packet of energy that has a particular frequency and travels through space.
- **Light** is **electromagnetic radiation**.
- The color of **light** depends on its wavelength. Red light has a longer wavelength, and violet light has a shorter wavelength.
- The colors in fireworks are caused by metal atoms releasing **photons**.

12.6 Vocabulary

Basic Words

emit (ee-MIHT)

light (LIYT)

photon (PHOH-tahn)

sonic (SAH-nihk)

Challenge Words

electromagnetic (ih-lehk-troh-maag-NEH-tihk)

particle (PAHR-tih-kuhl)

quantum mechanics (KWAHN-tuhm mih-KAA-nihks)

radiation (ray-dee-AY-shuhn)

special relativity (SPEH-shuhl reh-luh-TIH-vuh-tee)

visible (VIH-zuh-buhl)

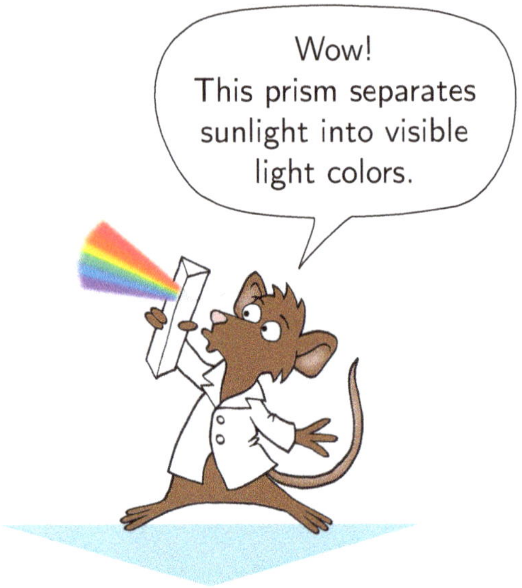

CHAPTER 13

Mirror MIRROR

13.1 Introduction

What happens when you look into a mirror? You know that you will see an image of yourself. If you move your arms up, you can see how your arms move. If you squish your face or stick out your tongue, you can see how your face moves and what your tongue looks like in the mirror. But what exactly is going on when you look into a mirror?

One clue that light has something to do with it is the fact that you can't see yourself in the mirror if you are in a completely dark room. You know you still exist, but no matter how hard you try, if there is no light in the room, you can't see your image in the mirror.

As we saw in the last few chapters, light behaves like both a particle and a wave. The particle-like nature of light gives us photographs, and the wave-like nature of light gives us images of ourselves when we look into mirrors.

Although some objects **emit** their own light, most of the objects we see around us are visible because they **reflect** light that has come from another source. When you turn on a lamp in a dark room, the light from the bulb rushes out at the speed of light to illuminate the objects in the room.

As we saw in Chapter 11, sound waves can interact with objects in several different ways. Sound waves can be reflected, transmitted, absorbed, and refracted. The same is true with light waves.

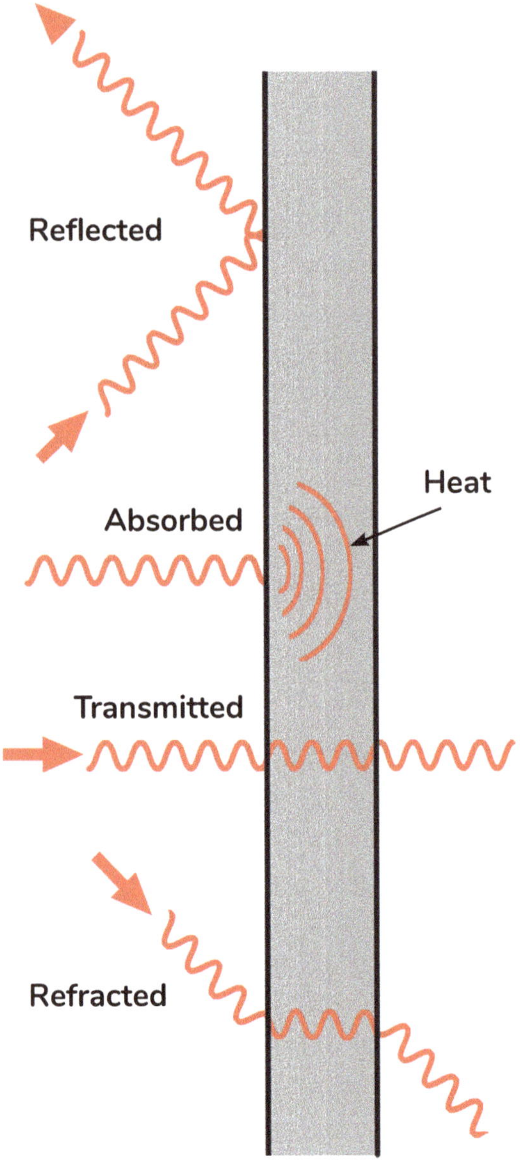

When a beam of light hits an object, the light can be **reflected** by the material it contacts, **absorbed** by the material and turned into heat, pass through the material completely (be **transmitted**), or it can be bent by the material (be **refracted**).

Whether a beam of light is reflected, absorbed, transmitted, or refracted depends partly on the type of material the beam of light encounters and partly on the wavelength of the light beam.

What determines how light interacts with different materials? The pages of this text are visible because light from the Sun or a lamp is interacting with the atoms on the page. The white part of the page is reflecting all the visible wavelengths of light and therefore appears white. The black ink is absorbing all the visible wavelengths and therefore appears black. Each colored part of the page is absorbing all the wavelengths except the wavelength that produces the color you see, which is reflected rather than absorbed. So the blue colors on a page are reflecting blue wavelengths, the red colors are reflecting red wavelengths, and the yellow colors are reflecting yellow wavelengths. Different parts of the page are interacting with light differently.

13.2 Blue Skies and Red Sunsets

Why is the sky blue and why are sunrises and sunsets multiple colors?

You know that when light hits an object, it can be reflected back. When light interacts with molecules in the air, the light is reflected in all possible directions. This is called **scattering**. It turns out that for very small particles like air molecules, light with shorter wavelengths (higher frequencies) is scattered much more strongly than light with longer wavelengths. So blue light is scattered more strongly than red light. The sky is blue because the light you see in the sky that comes from the Sun is scattered by air molecules.

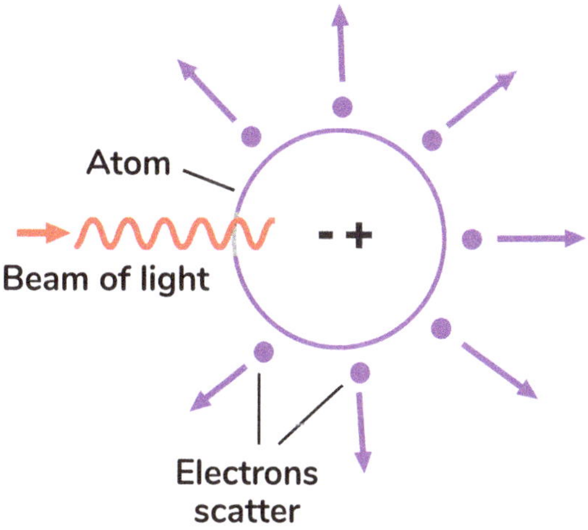

Violet is scattered the most, but our eyes see blue wavelengths more easily than the shorter violet wavelengths, so the sky looks blue to our eyes.

In places like New Mexico and other desert areas where the humidity is low and the air is clear of dust and particles from pollution, the sky is a brilliant, beautiful blue. In places where the humidity is high and there are lots of impurities in the air, the sky will be less blue because it will contain some white coloring due to large particles in the air scattering all the colors of the visible spectrum.

If you've ever been to New Mexico, most evenings you can watch the Sun change color from yellow to a beautiful orange or red as it sets. If there are clouds in the sky, the clouds can turn brilliant pinks, oranges, and reds. When the Sun is setting, sunlight travels through the thickest part of the atmosphere and more of the short blue and violet wavelengths are scattered away from your eyes. Because red and orange wavelengths are scattered the least, they are transmitted through the atmosphere rather than scattered. This makes the Sun appear more orange and red as it sets.

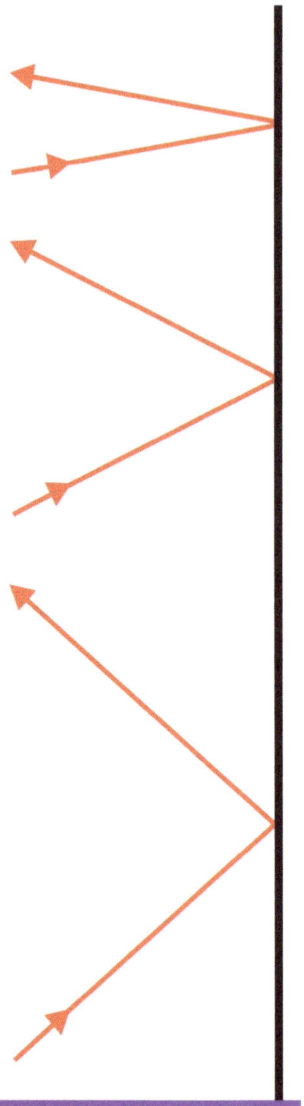
Examples of Law of Reflection

13.3 Reflection

When you look into a mirror, does your image get scattered? Do you come out looking like a bunch of blue, green, red, yellow, and purple dots? No! You do not look like a bunch of colored dots because the light hitting the smooth surface of a mirror is being reflected by the mirror and not scattered in many directions.

When light is reflected, it obeys the **law of reflection.** The law of reflection states that the angle of reflected light will be equal to the angle of the incoming light. This means that a beam of light will be reflected at exactly the same angle as it came in and not reflected (or scattered) at all different angles in all directions.

When you look into a flat mirror you see a duplicate of yourself. You are the same height, the same width, and all of your limbs are the correct proportion. What happens when you walk close to the mirror? Does your image appear to walk with you? What happens if you back away from the mirror? Does your image also back away? Your image follows you wherever you go because the light being reflected obeys the law of reflection. No matter where you move, the angle of reflected light equals the angle of incoming light and your image follows you.

Have you ever been to a fun house and noticed that one mirror made you look narrower and another mirror made you look wider? A curved mirror will change the way you see yourself. Because light that reflects from a smooth surface must obey the law of reflection, if the mirror is curved, the image you see will be distorted.

If the mirror is curved with the edges on the sides going back and the middle coming toward you, the image you see will be smaller, or thinner. If the mirror is curved with the side edges coming toward you and the middle going away from you, the image you see will be larger, or thicker.

13.4 Refraction

What happens when you look at your fishing line as it enters the water? Does it appear to bend? What happens if you take a straight stick or a pole and put it in the water? Does it look crooked? Are the fishing line and the stick actually bending, or do they just *look* like they are bending?

The fishing line and the stick are not really bending. It is the light being reflected back to your eyes that is bending, and this makes the objects in the water look bent.

When light encounters a medium like water, glass, or oil, not only do the light rays bend, but the speed of light also changes. The speed of light in a vacuum is 299,790 kilometers per second, but it slows down when it bumps into the atoms and molecules in different materials. The speed of light in the atmosphere is just slightly less than in a vacuum. However, the speed of light in water is about 225,000 kilometers per second and in glass it is about 200,000 kilometers per second. In a diamond the speed of light travels less than 150,000 kilometers per second. But oddly enough, when light emerges on the other side of the material, it goes back to traveling at its original speed.

Because light bends when it encounters materials like glass and plastic, lenses can be ground or molded so that they bend light in a way that an image is formed. Lenses are used in cameras, telescopes, microscopes, and eyeglasses.

13.5 Types of Lenses

There are many different ways to create lenses. Lenses can be thick or thin, long or short, round, square, or oval. But regardless of how the lens is shaped, there are two main types of lenses — **converging lenses** and **diverging lenses**.

A **converging lens** bends the incoming light so that the light rays that come out of the opposite side of the lens **converge**, or meet, at a single point. The point where the light rays converge is called the **focal point**. Converging

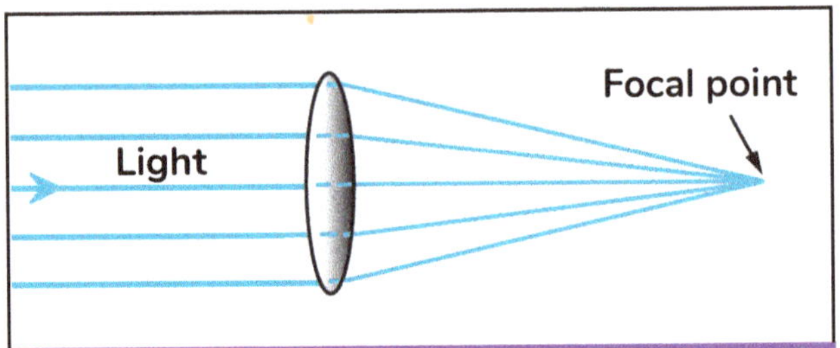

Light passing through a converging lens

lenses are thick in the middle and thin on the edges. The varying thickness allows each section of the lens to bend the light at a different angle. Parallel light rays coming into the lens are bent inward by the curvature of a convergent lens.

A **diverging lens** bends the incoming light so that the light rays that come out of the opposite side of the lens diverge, spreading the light rays away from each other. Divergent lenses are thin in the middle and thick on the edges. Again, this allows each section of the lens to bend the light at a different angle. Parallel light rays are bent outward by the curvature of a divergent lens.

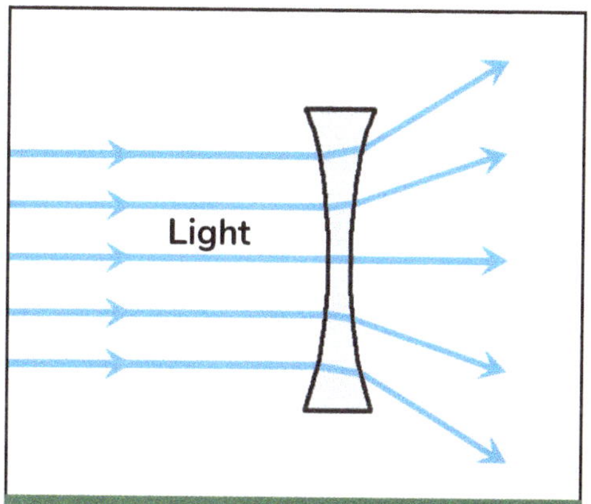

Light passing through a diverging lens

13.6 Summary

- When light hits an object it can be **reflected**, **transmitted**, **absorbed**, or **refracted**.
- When light is **reflected** in all possible directions, it is called **scattering**.
- When light is **reflected** and **does not scatter**, it obeys the **law of reflection** which states that the angle of incoming light is the same as the angle of outgoing light.
- When light passes through a material that bends the light rays, it is called **refraction**.
- Lenses use **refraction** to **converge** light onto a **focal point** or **diverge** light away from the center.

13.7 Vocabulary

Basic Words

absorb (uhb-SAWRB)

focal (FOH-kuhl)

lens (LEHNZ)

reflect (rih-FLEHKT)

refract (rih-FRAAKT)

transmit (traans-MIHT)

Challenge Words

converging (kuhn-VUHRJ-ihng)

diverging (duh-VUHRJ-ihng)

CHAPTER 14

What Goes AROUND?

14.1 Introduction

In previous books we have explored some of Earth's cycles, including the water cycle, the carbon-oxygen cycle, and the nitrogen cycle. In this chapter we'll take a closer look at three of Earth's major **geological cycles**, the **atmospheric cycle**, the **rock cycle**, and the **energy cycle**.

14.2 The Atmospheric Cycle

Recall that the air we breathe is part of Earth's **atmosphere**, and the atmosphere sits just above Earth's **crust**, extending for several miles above Earth's surface. Every day you participate in Earth's **atmospheric cycle** just by breathing. When you breathe, you inhale oxygen and exhale carbon dioxide. This **carbon-oxygen cycle** is critical for life on Earth and is part of the atmospheric cycle. The atmospheric cycle also includes **weather** and **climate**.

When you get caught in a torrential downpour only to have the Sun come out and dry you off a few minutes later, you are experiencing **weather**. Weather is the part of the atmospheric cycle that refers to daily changes in sunshine, temperature, rainfall, and wind.

Weather can vary from day to day and during the day. Depending on where you live and the time of year, you might need to be prepared for a sunny day that turns into a snowy blizzard, or a cold, rainy morning that turns into a hot, sunny afternoon. You might start out the day with the Sun shining brightly and feel a cool breeze on your skin. But by midmorning, clouds may cover the sky, and later in the day you might find yourself in a torrential downpour. These are all changes in the weather.

While weather refers to day to day changes, **climate** refers to the temperature, **precipitation**, wind, amount of sunshine, and other factors in a certain region averaged over a long period of time. The climate differs from one region of Earth to another. For example, in some parts of the world, the overall climate is cold and wet, even though some days may be hot and dry. In other parts of the world the overall climate is hot and dry even though some days may be cold and wet.

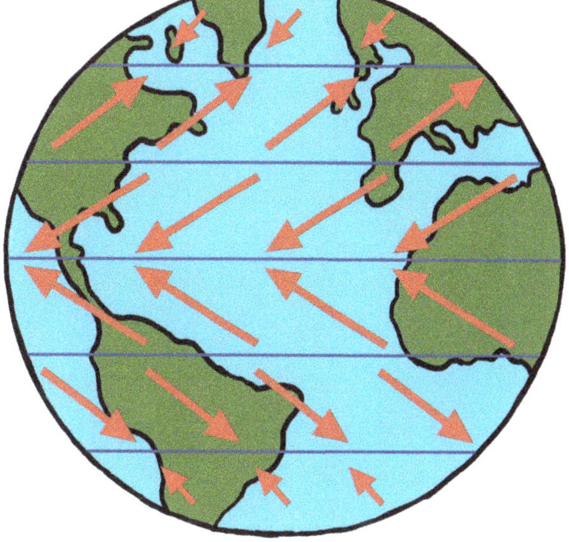

The atmosphere moves around the Earth as large and small continuously changing rivers of air. These rivers of air cover the entire surface of Earth. The movement of these rivers of air is powered by heat from the Sun and by Earth's rotation.

The reason weather varies from place to place is due to differences in **atmospheric pressure** (also called **air pressure**). Atmospheric pressure is the force exerted on us by the weight of air molecules in the Earth's atmosphere. **Pressure** refers to the measure of the number of molecules and how fast they are moving in a given space. For example, when you blow up a balloon, you are filling it with air. As you blow more and more air molecules into the balloon, the balloon gets bigger, the molecules are packed closer together, and there is more pressure from air molecules pushing against the sides of the balloon. If you let air out of the balloon, the pressure inside the balloon decreases as molecules leave the balloon.

What happens if you heat or cool a balloon? Heating a balloon will cause the air molecules to move faster and push harder on the sides of the balloon. This increases the pressure. If you put the balloon in the freezer, the balloon will start to shrink because the air molecules slow down, reducing the pressure.

Because it takes more energy to maintain high pressure, given the choice, air will move from high pressure to lower pressure. When you let the air out of the balloon, the air will rush out from the high pressure area inside the balloon to the lower pressure area outside the balloon.

The same thing happens on a large scale in the atmosphere. Air molecules are trapped in the space between the surface of the Earth and the upper atmosphere. The number of air molecules in a certain region and how fast they are moving is constantly changing because of heat from the Sun and heating and cooling from the oceans and land. Some regions have higher pressures with more air molecules, and other regions will have lower pressures with fewer air molecules. When a low pressure area develops, air from the surrounding higher pressure region will rush in to fill it up, creating storms. Low pressure areas tend to have lots of cloud cover and rain as storms move in, and high pressure areas are sunny and clear.

The atmospheric cycle is coupled very closely to the water cycle. Recall that in the water cycle, water moves from Earth's rivers and oceans to the atmosphere and then is delivered back to Earth's surface as rain, snow, and other types of precipitation. For example, storms that begin over the oceans, such as tropical storms, hurricanes, and typhoons, can deliver massive amounts of water to coastlines.

El Niño is a good example of both the atmospheric and water cycles working together. El Niño is a cyclical weather pattern that happens every four to seven years. El Niño usually causes severe storms and flooding along the western coastline of the Americas and often brings much needed relief from drought in the southwestern portion of North America. An El Niño event will generally begin in late December with warm Pacific

Ocean surface water traveling eastward. When this happens, the normal atmospheric wind patterns switch direction, bringing rains to the normally dry western South American and North American coasts. Eventually the pattern switches again and the cycle repeats.

14.3 The Rock Cycle

Rocks are continuously created, transformed, broken down, and created again. This is called the **rock cycle**. **Igneous rocks, sedimentary rocks,** and **metamorphic rocks** are all formed and shaped by the heating and cooling of minerals and by pressure, weather, and ocean activity. Rocks are continuously being broken down and moved around by the forces of wind and water.

For example, you might live in a house on the beach and notice that as the waves roll in and move back out, the sand on the beach shifts. You might notice that in the winter big waves take away large amounts of sand and then deposit the sand on offshore sandbars. In the summer, sand is moved back to the beach by smaller waves. Or you might observe how very large storms can destroy a beach and sometimes change the beach front so drastically that homes need to be removed or rebuilt. The movement of sand by water changes the landscape and is part of the rock cycle.

The rock cycle happens slowly in many areas, but where there is volcanic activity, such as in the Hawaiian islands, you can watch it occurring daily. Rocks being formed by volcanic eruptions is the most spectacular of all rock-forming events.

Basalt is the most common type of volcanic rock. It is dark and even-textured and makes up most of the rock formation on the Hawaiian islands. When volcanic rock forms, you can watch red hot lava flow from the volcanic source, forming basalt as it cools. When lava contacts the cool ocean, jets of steam and hot water spray out everywhere as the water boils. This often creates a spongy-looking type of rock called **pumice** that is very light because it is full of air holes.

Once volcanic rock has cooled and hardened, it is exposed to **weathering**, and a much slower process in the rock cycle begins. The actions of weather, including wind and rain, can break off tiny pieces, or grains, of rock. The grains of rock then get washed away by rivers or blown by winds and moved into the seas. Beaches contain billions and billions of tiny grains of rock that have been broken off bigger rocks. When these tiny grains get into the oceans, they sink to the ocean floor, and over time, layers and layers of grains of rock begin accumulating. As these grains accumulate and form many layers, pressure and heat are created from the weight of the upper layers. The pressure and heat act like a kind of glue, sticking the sediments together and creating sedimentary rocks.

Many different types of **sedimentary rocks** can be seen along the winding roads of the North American Southwest. They are easy to spot, looking like a layered cake, and can have sharply contrasting red, white, brown, and black layers. These layers formed at the bottom of an ocean over very long periods of time, and they contain sandstones, shales, mudstones, and limestones. Their presence on the top of mesas and other landforms shows that the rock cycle actively transforms Earth's crust. Rock that was once at the bottom of the ocean is now at the top of a mountain, and eventually the top of the mountain wears away and goes back into the ocean, carried by rivers and streams. There it is formed into rock again.

Sometimes sedimentary rocks are not pushed upward to form mountains but instead are buried deep within the Earth's crust. When this happens, intense heat and pressure transform them into metamorphic rock.

One of the most beautiful **metamorphic rock** is marble. Marble starts out primarily as the skeletal remains of small ocean animals. Over time, these skeletal remains turn into limestone, a sedimentary rock that is then further crushed by layers and layers of sediments. The layers of sediments create intense heat and pressure, turning the limestone into marble. When **tectonic plates** collide, these marble layers are thrust upward and can be removed by humans for art and architecture, and when the marble is exposed to the weather, it is gradually worn away.

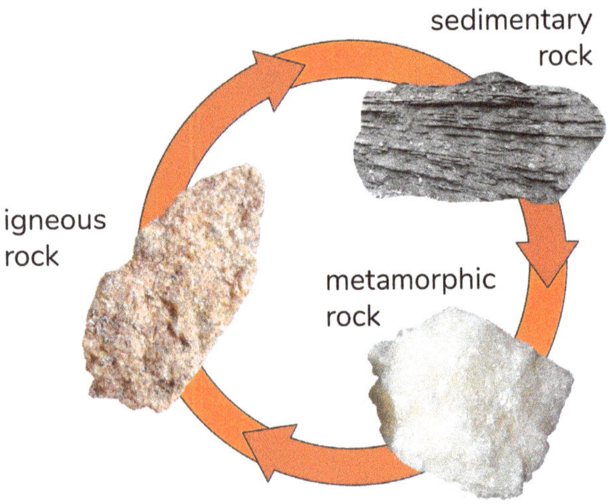

Igneous, sedimentary, and metamorphic rocks all play a role in the rock cycle. Volcanic eruptions create igneous rocks that can be weathered and turned into sedimentary rocks. Sedimentary rocks, in turn, can be transformed into metamorphic rocks that are themselves weathered and converted back to sedimentary rocks. All of these rocks can also sink deep into the Earth's surface, melt, and be reformed into new igneous rocks, or they may return to the surface as lava.

14.4 The Energy Cycle

One of the most important cycles to observe and understand is Earth's **energy cycle**. Energy drives all geological, atmospheric, biological, and chemical cycles. Without energy from deep within Earth, volcanoes could not erupt and create rocks and mountains, and massive tectonic plates would not collide. Without energy coming from the Sun, weather would not change and plant life would not be able to create food for all the living things on Earth.

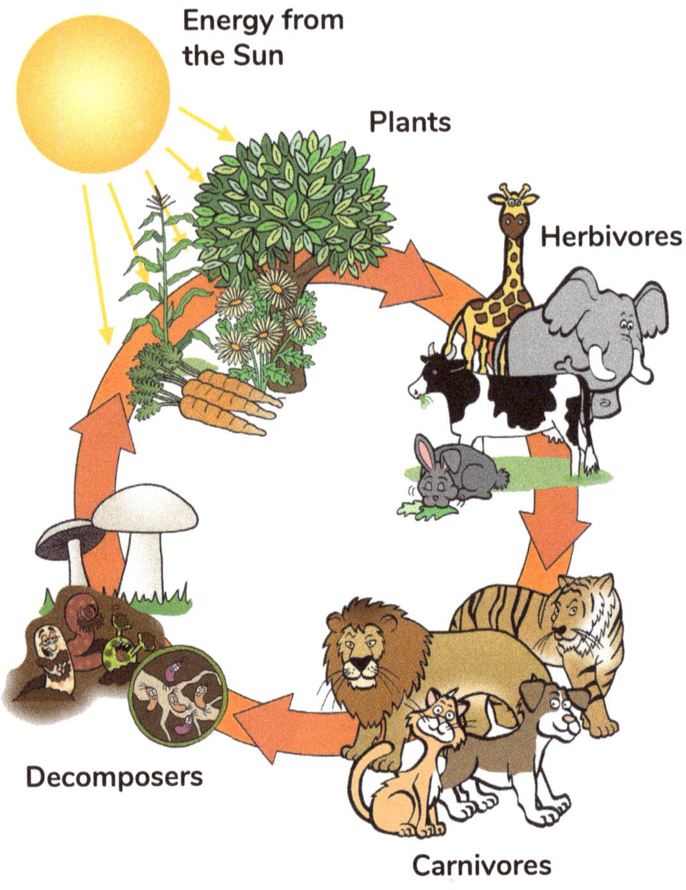

Although some surface energy comes from inside the Earth, the majority of Earth's energy comes from the Sun. The Sun is the source of all the energy needed by plants to make food.

In order for Earth's temperature to remain balanced, the amount of incoming heat needs to be equal to the amount of heat leaving the surface. If the amount of incoming and outgoing heat was not balanced, Earth would get progressively hotter or colder over time. The balance between incoming and outgoing heat is known as Earth's **heat budget.**

Although the Sun produces massive amounts of light and heat energy, the Earth only receives about 2 billionths of the Sun's total energy output. Of this 2 billionths about 34% is reflected back out to space by clouds, land, and snow and glaciers on the ground. This reflective action is called the Earth's **albedo**. The word albedo comes from the Latin word *albus* which means "white." Because white surfaces reflect light, albedo refers to the amount of light reflected by a surface and in particular the amount of light reflected by a celestial body such as a planet.

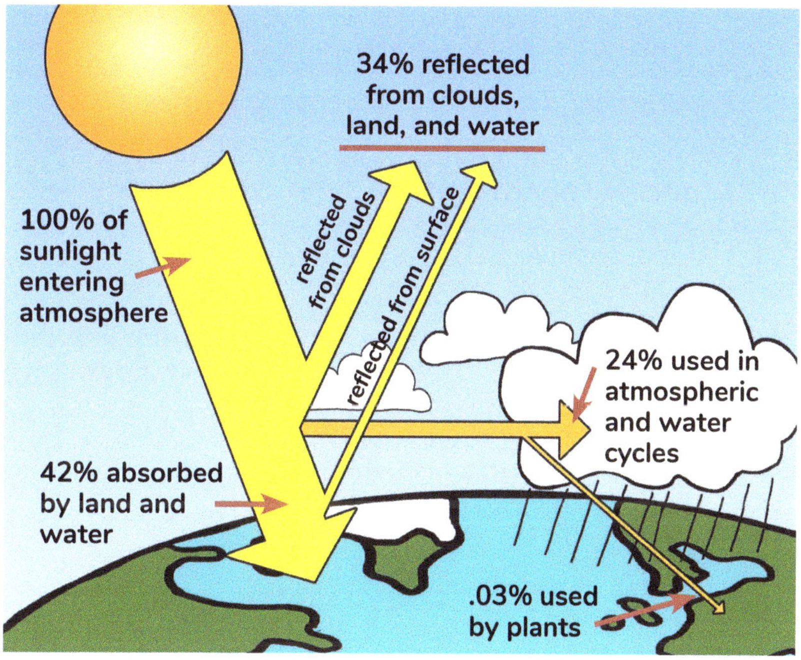

The Sun also warms Earth's surface, and about 42% of the 2 billionths of the Sun's energy is absorbed by land and water. The water and atmospheric cycles use about 24% of the Sun's energy to move more than 495,00 cubic kilometers (119,000 cubic miles) of water vapor though the atmosphere. The remaining fraction, about 0.03%, is used by plants for **photosynthesis**.

Plants convert solar energy in the form of light into chemical energy in the form of food. This chemical energy is consumed by plant-eating organisms, like cows, that are themselves eaten by other organisms, like humans or lions. Eventually the animals die and the chemical energy stored in their bodies is converted back into a form utilized by plants. The chemicals reenter the energy cycle and the process repeats. It is astonishing to consider that just a small fraction of the Sun's energy fuels all living things on Earth.

14.5 Summary

- The **atmospheric cycle** is responsible for both **weather** and **climate**.
- The **atmospheric cycle** is coupled to both the **water cycle** and the **energy cycle**.
- Changes in **weather** occur when changes in **atmospheric pressure** occur.
- Earth's rocks are continuously created, transformed, destroyed, and created again during the **rock cycle.**
- **Igneous**, **sedimentary**, and **metamorphic rocks** all play a role in the **rock cycle**.
- Energy from the Sun is reflected and absorbed by Earth's surface.
- A small percentage of the Sun's energy enters the **energy cycle** and is used by plants for photosynthesis.

14.6 Vocabulary

Basic Words

basalt (buh-SAWLT)

budget (BUH-juht)

climate (KLIY-muht)

cycle (SIY-kuhl)

El Niño (EHL NEE-nyoh)

pressure (PREH-shuhr)

pumice (PUH-muhs)

weather (WEH-thuhr)

Challenge Words

albedo (aal-BEE-doh)

atmosphere (AAT-muh-sfihr)

atmospheric (aat-muh-SFIHR-ihk)

geological (jee-uh-LAH-jih-kuhl)

igneous (IHG-nee-uhs)

metamorphic (meh-tuh-MAWR-fihk)

photosynthesis (foh-toh-SIHN-thuh-sihs)

sedimentary (seh-duh-MEHN-tuh-ree)

tectonic (tehk-TAH-nihk)

CHAPTER 15

What Is ECOLOGY?

15.1 Introduction

What do you see in your neighborhood? If you live in the city, your neighborhood might be mostly made of paved roads, brick buildings, and concrete sidewalks. But if you look closely, even in the city you can find plants, trees, bugs, birds, cats, and dogs. If you live in the country, your neighborhood may be filled with large stretches of grassy fields, trees, and perhaps cows, goats, sheep, or horses.

Both city and country neighborhoods have an **ecology**. The word ecology comes from the Greek words *oikos* which means "house, dwelling place, habitation" and *logia* which means "study of." The word ecology means both the relationships between living things and their **environment** and the study of these relationships.

Scientists who study ecology look closely at how living things interact with their environment and how changes to the environment impact and change the way organisms live. Scientists study how plants, animals, and microscopic organisms interact with the Sun, Earth, soil, atmosphere, weather, water, and energy. Together, both living organisms and nonliving influences create an **ecological system,** or **ecosystem**.

15.2 Parts of an Ecosystem

Ecosystems can be as large as the Atlantic Ocean or as small as a rain puddle. Ecosystems occur on every surface of land and in every body of water. Although all ecosystems are different from each other, they share a few common characteristics.

All ecosystems contain living things. All of the living things in an ecosystem form an **ecological community** that is made up of individual organisms that interact with each other to maintain life.

For example, an oceanic ecosystem may have an ecological community that includes bony fish, corals, jellyfish, sharks, sea slugs, shrimp, and tiny plankton. All of these organisms interact with each other to support life. Large fish may prey on smaller fish, and sharks may prey on large fish. However, small fish may eat the decaying flesh of dead large fish and sharks. Corals may provide shelter for small fish, and small crabs, like the hermit crab, may use previously occupied snail shells as a portable home.

A completely different ecosystem will have a completely different set of living things interacting to maintain life. For example, in a forest ecosystem the ecological community may contain pine trees, spiders, ants, hummingbirds, bobcats, snakes, gray tailed squirrels, raspberry bushes, and hundreds of mushrooms. Raspberries provide food for squirrels, squirrels provide food for bobcats, and dead animals provide food for mushrooms. All of these living things work together to create a community that supports life.

All ecosystems are also made of nonliving things. Ecosystems can contain rocks, soil, water, sunlight, oxygen, heat, rainfall, snow, or salty ocean water.

Living things interact and affect nonliving things, and nonliving things interact and affect living things. For example, when water currents, temperatures, and food levels change in certain water ecosystems, an overpopulation of **algae** may result. This overgrowth of algae is called an **algal bloom**. Algal blooms can occur in both freshwater and saltwater ecosystems and can be either harmless or harmful to the

surrounding environment. Harmful algal blooms can kill large numbers of fish and shellfish, change the nutrient and oxygen content in the water, and destroy aquatic plant life.

Algal blooms can occur as a result of natural causes, but some occur as a result of human activities. For example, some algal blooms occur when an excess of nitrates and phosphates enter the water system from agricultural processes.

Knowing how living things affect an ecosystem and how changes in an ecosystem affect living things is an important part of the science of ecology.

15.3 Trophic Levels

In Chapter 14 we explored how the flow of energy from the Sun to plants, from plants to animals, and then from animals back to plants is part of the **energy cycle**. This flow of energy is found in all ecosystems.

Each ecosystem has a **food chain**. A food chain describes the order in which the organisms in an ecosystem use each other as a food source. Plants are at the first level of the food chain and herbivorous animals that eat the plants are next. Carnivorous animals that eat other animals are at the top of the food chain.

A food chain can be broken down into **trophic levels**. A trophic level is the position an organism occupies in the food chain.

1. The first trophic level includes all the plants and other organisms that make food directly from the Sun's energy. Organisms in the first trophic level are called **producers**.

2. The second trophic level includes all the **herbivores** — the animals that eat only plants. For example, cows, rabbits, deer, and buffalo are all in the second trophic level because they all eat only plants

for food. All of the organisms in the second trophic level are **primary consumers**.

3. The third trophic level includes all **carnivores** that eat herbivores. For example, cheetahs, hyenas, moles, birds, foxes, and weasels eat other animals and are called **secondary consumers**.

4. The fourth trophic level includes all carnivores that eat both herbivores and other carnivores. For example, eagles, hawks, and alligators eat both herbivores and other carnivores. When organisms in this trophic level eat animals that are already dead, they are called **scavengers**.

5. The **apex predators** are in the fifth trophic level at the top of the food chain. These organisms have no other predators. For example, tigers are the largest predator in the world that does not hunt in water, and tigers have no other predators

Food chain

except humans. Killer whales are apex predators that live in the ocean and kill both sea lions and other whales. Humans are apex predators and are at the top of the food chain in much of the world.

6. Organisms that consume dead plants and animals are called **decomposers**. Decomposers get their energy from all the trophic levels and include organisms such as mushrooms and bacteria.

It is important to consider that only a small portion of the energy in one trophic level is transferred to a higher trohpic level. In fact, only about 10% of the energy available at one trophic level reaches the next trophic level. This means that a smaller and smaller amount of energy is available to organisms as the energy travels up the food chain. Plants must constantly make food from the Sun's energy in order for energy to reach the upper trophic levels. Without the Sun the food chain and all its trophic levels would collapse.

15.4 Ecological Niches

Every organism occupies a particular place within an ecosystem called its **ecological niche**. Every organism has a particular way to survive and use matter and energy. In other words, the ecological niche of an organism is how it makes its living.

The word **niche** comes from the Middle French word *nicher*, which means "to nest." This term was first used by the naturalist **Joseph Grinnell** in 1917 to describe all the factors needed by an organism to survive in a particular habitat.

Organisms within a niche need the right conditions in order to grow, and they depend on each other for survival — they are **interdependent**. Plants need the right amount of moisture and the right kind of soil in order to grow in a particular area. Herbivores depend on having enough of the right kinds of plants to eat. Carnivores need enough of the right kinds of herbivores to eat. Predators keep the animals they prey on from overpopulating an area, which could lead to starvation.

Organisms in a niche vary in population depending on how they respond to food sources, predators, and physical changes in climate, soil, and other factors. Since organisms need specific conditions in order to survive, different plants and animals live in different ecological niches.

For example, polar bears would not survive in tropical climates. Polar bears have thick fur coats that are perfect for living where temperatures are very cold but would make them much too hot in tropical climates. Also, polar bears need to live where there are seals and other cold water marine animals to eat.

The American Flamingo, found in the Caribbean islands and Central America, would not survive in the cold climates found near the North Pole. Flamingos eat blue-green algae and small brine shrimp that need warm temperatures to grow. Flamingos have long, thin legs that allow them to wade in water where they use their feet to stir up food from the warm mud at the bottom.

15.5 Law of Unintended Consequences

What happens if you introduce a new fish to an established ecological niche? In the 1950s in an effort to increase the commercial fishing market, a species of large fish, the **Nile perch**, was introduced to **Lake Victoria**, the largest tropical lake and second largest freshwater lake in the world. Lake Victoria sits on the borders of Kenya, Uganda, and Tanzania in East Africa.

In the 1970s scientists observed a sudden decrease in the population of native fish called **cichlids**. Cichlids are small, bony fish that eat dead and decaying matter that floats on the top of the lake. Because the cichlids no longer consumed this decaying matter, it sank to the bottom of the lake and used large amounts of oxygen in the process of decaying. The loss of oxygen in the water, particularly in the deepest parts of the lake, created dead zones where fish and plants cannot grow.

Also, the cichlids helped control the algae population. As the number of cichlids declined, unchecked algal blooms spread across Lake Victoria, and large quantities of dead algae sank. The decaying of large amounts of algae increased the loss of oxygen in certain parts of the lake.

The local people depended on the cichlids for food and did not have the big boats needed to catch the Nile perch. With their food source destroyed, many of the native people were not able to get enough food to eat and had to change the way they were living. Over time, the introduction of one species drastically altered the ecosystem of Lake Victoria.

Ecosystems are complex and have many parts that work together and interact with each other in ways that people often do not understand. When a part of a complex system is changed, the results can be unexpected and undesirable. This is referred to as the **law of unintended consequences** — when an unexpected outcome occurs due to a change that is made.

One of the most important concerns of ecology is understanding how changing one thing, whether it is a species, a chemical, a temperature, or a food source, makes something else happen — often in ways we are not able to predict. By understanding how ecosystems work and how their many parts work together, and by keeping the law of unintended consequences in mind, people can make changes that help ecosystems thrive.

15.6 Summary

- **Ecology** is the study of how living things interact with other living things and their **environment**.
- Together, both living organisms and nonliving influences create an **ecological system** or **ecosystem**.
- There are several **trophic levels** through which energy flows from the Sun all the way to the **decomposers**.
- An **ecological niche** is the place within an ecosystem that has the right conditions for a particular organism to live.
- Small changes can drastically alter an entire **ecosystem**.

15.7 Vocabulary

Basic Words

algae (AAL-jee)

algal (AAL-guhl)

cichlid (SIH-kluhd)

Joseph Grinnell (JOH-zuhf grih-NEHL)

niche (NIHCH or NEESH)

trophic (TROH-fihk)

Challenge Words

apex predator (AY-pehks PREH-duh-tuhr)

carnivore (KAHR-nuh-vawr)

decomposer (dee-kuhm-POH-zuhr)

ecological (ee-kuh-LAH-jih-kuhl)

ecosystem (EE-koh-sih-stuhm)

ecology (ih-KAH-luh-jee)

environment (ihn-VIY-ruhn-muhnt)

herbivore (UHR-buh-vawr)

interdependent (ihn-tuhr-dih-PEHN-duhnt)

producer (pruh-DOO-suhr)

scavenger (SKAA-vuhn-juhr)

unintended consequence (uhn-ihn-TEHN-duhd KAHN-suh-kwehns)

CHAPTER 16

Natural RESOURCES

16.1 Introduction

Unlike other living things, humans use materials from the Earth to drastically change their environment. We dig deep in the Earth's surface for metals, oils, and salts and convert them into automobiles, gasoline, and food. We build dams for hydroelectric power and remove mountains to make roadbeds. As our numbers grow, we continually search the globe for new sources of minerals, coal, and uranium. As our technology advances, we use more sophisticated instruments and combine geology, chemistry, and physics to extract more raw materials from the Earth and more food and energy from these raw materials.

As the human population continues to increase, our needs for food and energy grow. The term **natural resources** refers to the materials found in and on Earth in a natural state. Natural resources include water, air, plants, animals, sunlight, crude oil, natural gas, coal, uranium, salt, and a variety of minerals. Humans convert these natural resources into usable forms of food and energy. One of the most important tasks for scientists and engineers is to figure out how to convert natural resources to food and energy without wasting or depleting them.

16.2 Oil, Gas, and Coal

Oil, natural gas, and coal are called **fossil fuels**. Fossil fuels come from dead plant and animal matter and are made mostly of long chains and rings of carbon and hydrogen, called **hydrocarbons**.

Oil is used to power cars, homes, and airplanes. Oil as it comes directly from the ground is called **crude oil**. Crude oil is purified by chemical and physical processes into various types of oil and gasoline.

The use of oil as a natural resource goes all the way back to antiquity. For example, early Chinese people drilled for oil using a piece of iron attached to a bamboo pole, and in Roman times crude oil was used as tar for paving roads and for heating and lighting.

Oil is made of a variety of hydrocarbons and these hydrocarbons can be **refined**, or separated from each other, by the chemical process of **distillation**. To **distill** crude oil, large oil refineries heat the crude oil to very high temperatures. When the crude oil is heated, lighter, smaller hydrocarbons are separated from heavier, larger hydrocarbons. Once

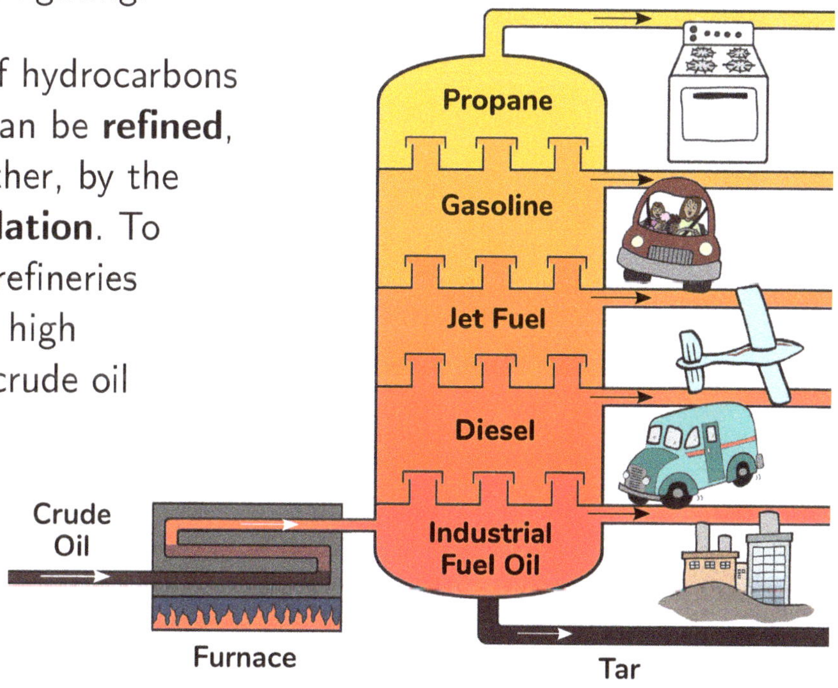

the different hydrocarbons are separated, they can be collected as different **fractions**. A fraction is a group of hydrocarbons that have similar chemical properties. Diesel fuel can be separated from gasoline, and gasoline can be separated from kerosene and propane.

methane

propane

butane

Natural gas is a hydrocarbon that is also found in deep underground rock formations. It is sometimes found in the same reservoir as oil and sometimes by itself in its own reservoir. In many parts of the world, natural gas is used to heat homes, provide fuel for stoves, and generate electricity.

Natural gas is primarily made of methane. Methane is a small hydrocarbon made of one carbon and four hydrogens and is a gas at room temperature. Natural gas also contains other small hydrocarbons including ethane, propane, and butane and also some impurities such as water and carbon dioxide. Before natural gas can be used as a fuel, water and other impurities must be removed.

Coal was the first fossil fuel to be used industrially for energy. In North America, coal was used by Native Americans for cooking, heating, and baking pottery. In the early 1700s the first North American commercial coal mines were built. Coal was found to burn cleaner and hotter than charcoal made from wood and as a result was used to power the industrial revolution of the nineteenth century. Coal was used to run steam engines,

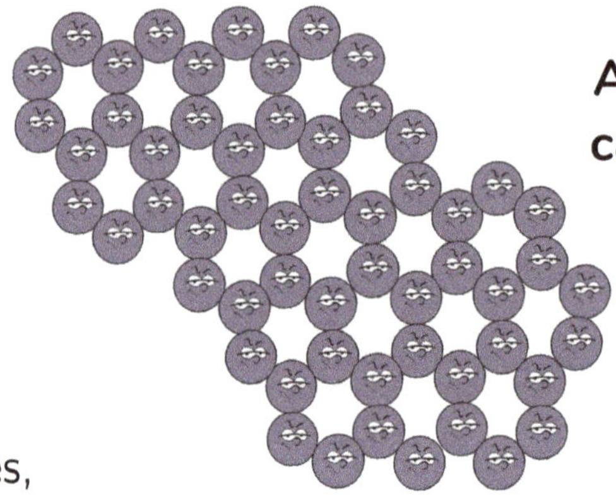
Arrangement of carbon atoms in coal (hydrogen not shown)

manufacture goods, power steamboats, and run railroad engines. According to the World Coal Association, today coal provides more than one-third of the world's electricity.

Coal is made of carbon and hydrogen atoms arranged in rings that look like a honeycomb.

By examining samples of different types of coal with a microscope, the English geologist **William Hutton** (1798-1860) found that all the samples contained plant cells. Since coal is made of plant **fossils**, it is a **fossil fuel**. The fossils in coal come from dead plants that were converted to minerals when they were subjected to extreme heat and pressure over very long periods of time. Coal is found in layers in sedimentary rock and can be a few to many meters thick.

16.3 Sun, Wind, and Water

Sun, wind, and water are considered to be **renewable fuels**. This means that, unlike fossil fuels which scientists believe could one day run out, sun, wind, and water power will not be **depleted**, or used up. Also, compared to the extraction of fossil fuels from Earth and the burning of them to get energy, renewable fuels cause less pollution.

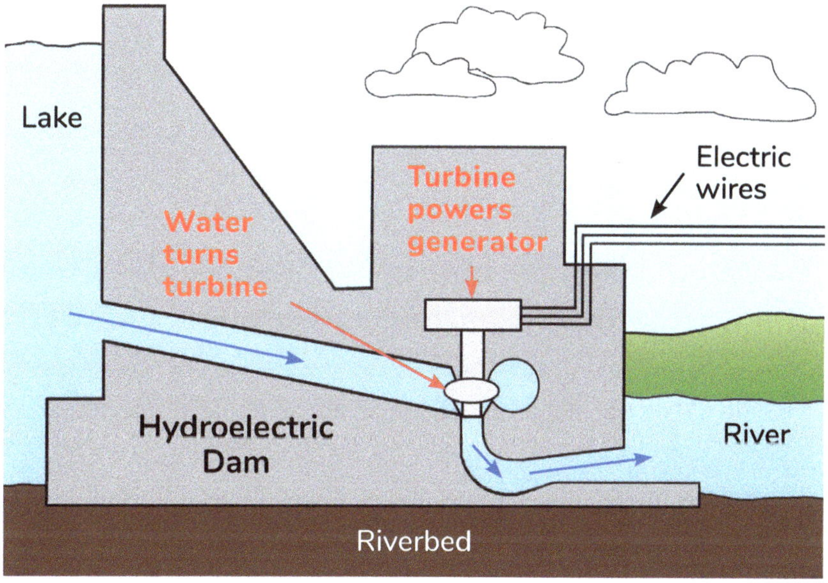

Renewable fuels are mainly used to generate electricity. In the late 1800s the first of these renewable fuels to be successfully converted to electricity in the United States was water. **Flowing water** can be captured by a dam and then pushed through a **turbine** to produce **hydroelectric power**. A turbine is a machine that converts the flow of liquid or air into work. In a dam, the turbines are connected to **generators** that

produce electricity. A generator is really a **motor** running in reverse. Motors convert electrical energy into mechanical energy, and a generator converts mechanical energy into electrical energy. The electricity is then transported to distant locations through electric cables.

Although hydroelectric power is considered a clean energy source because it does not produce harmful gases that escape into the environment, the production of hydroelectric power from rivers does modify the surrounding environment. When a river is dammed, a lake will form upstream. This can destroy plant, animal, and human habitats.

Another way to generate hydroelectric power is to use the movement of water in ocean tides to run turbines. **Tidal power** may be less harmful to the environment because there are no dams involved.

Solar power is generated when sunlight is collected and converted into electricity. **Solar panels** are made of thousands of tiny solar cells that release electrons when light energy strikes them. Solar cells are also called **photovoltaic cells**. The word photovoltaic comes from the Greek word *phos*, meaning "light," and *volt*, which is a unit of electrical energy named after **Alessandro Volta** (1745-1827 C.E.), the Italian physicist who invented the battery. Photovoltaic cells convert light energy directly into electricity.

Solar energy has been used to power small calculators, watches, cars, telescopes, and spacecraft. As the cost of creating photovoltaic cells continues to drop and as chemists and physicists perfect the conversion of sunlight to electricity with better and better materials, solar energy is becoming a competitive renewable energy source for many. Large solar panel fields can provide a city with electricity, and some individual families can generate enough electricity to power their homes and sell their excess electricity to utility companies.

Wind can also be converted into electricity. In the same way water flow is converted to hydroelectric energy, wind can turn a turbine and generate electricity.

People have been harnessing the wind to do work for centuries. The first windmills were used to grind grain and pump water. Early Persian, Greek, and Chinese people used windmills attached to grinding stones for converting grains into flour and to move water from one location to another. Both small and large windmills made of different materials including cloth, metal, and wood have been used in many different parts of the world.

The use of wind power to generate electricity was first developed in the late 1800s in Cleveland, Ohio by **Charles F. Brush** (1849-1929). The Brush windmill did not produce as much electricity as modern wind turbines, but was very successful in generating some electricity and operated for more than 20 years. Today, modern wind turbines can be seen in many countries, providing a clean alternative to fossil fuels.

16.4 Nuclear Power

Another way to generate energy is from **nuclear energy**. **Nuclear reactors** use **radioactive** material to generate electricity.

Nuclear energy is released in nuclear reactors. A **nuclear reaction** is different from a chemical reaction. In a nuclear reaction, the protons and neutrons get moved in and out of the nucleus, and the nucleus of the element changes. Therefore, the element itself changes. For example, in a nuclear reaction a nitrogen atom can absorb a neutron. When this happens, the nitrogen ejects a proton and turns into a special kind of carbon called **carbon-14.** In a chemical reaction the elements do not change, they just combine with other elements.

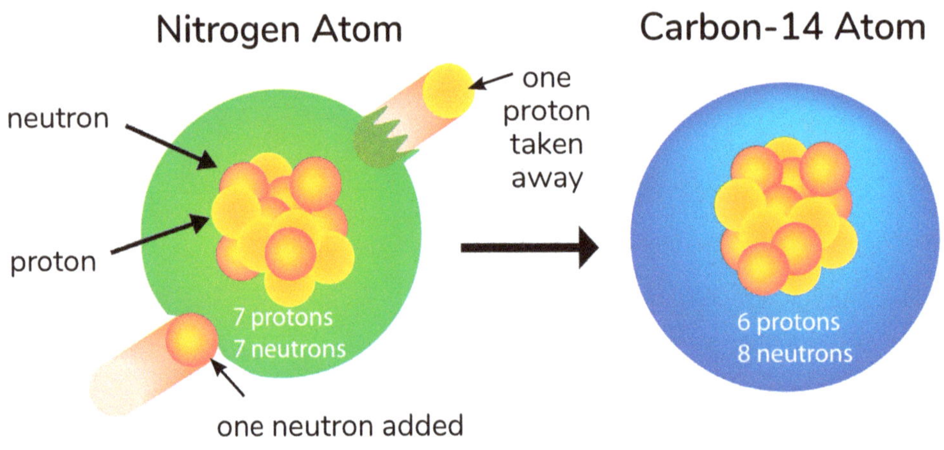

A nuclear reaction

When a uranium atom gets hit by a neutron, instead of losing just one neutron, it splits into other elements, such as krypton and barium. In the process, it gives off several neutrons. This process is called **nuclear fission**. Nuclear fission releases a huge amount of energy, much more than a chemical reaction. As little as 1 kilogram (2.2 pounds) of uranium fuel can produce as much energy as 2 metric tons (4,409 pounds) of coal. The neutrons generated by the fission reaction can hit more uranium atoms, each giving off more neutrons and causing a **chain reaction**. This chain reaction means that there are more and more neutrons available to combine with uranium atoms and generate electrical energy.

Nuclear Fission Reaction

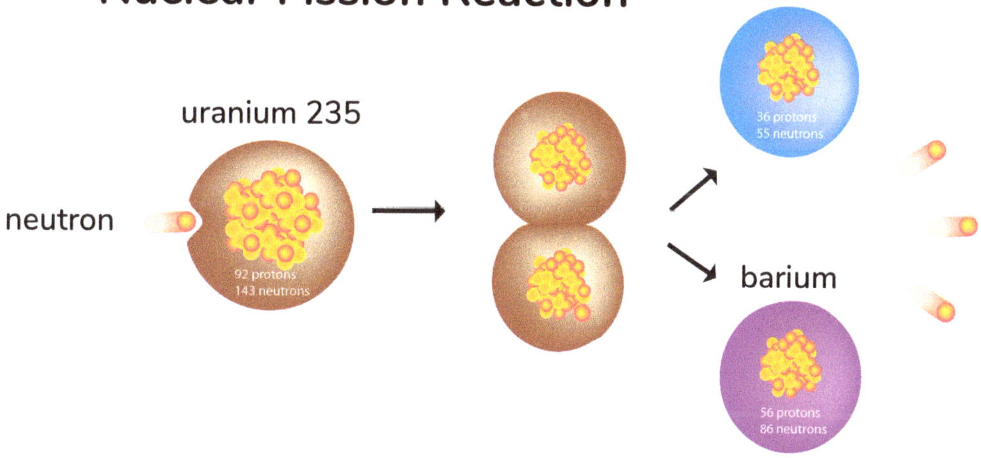

In a nuclear reactor, the heat generated by nuclear fission is used to heat water, which makes steam. The steam then turns a turbine that generates electricity. In this way, stored nuclear energy is converted to electrical energy.

Nuclear reactor

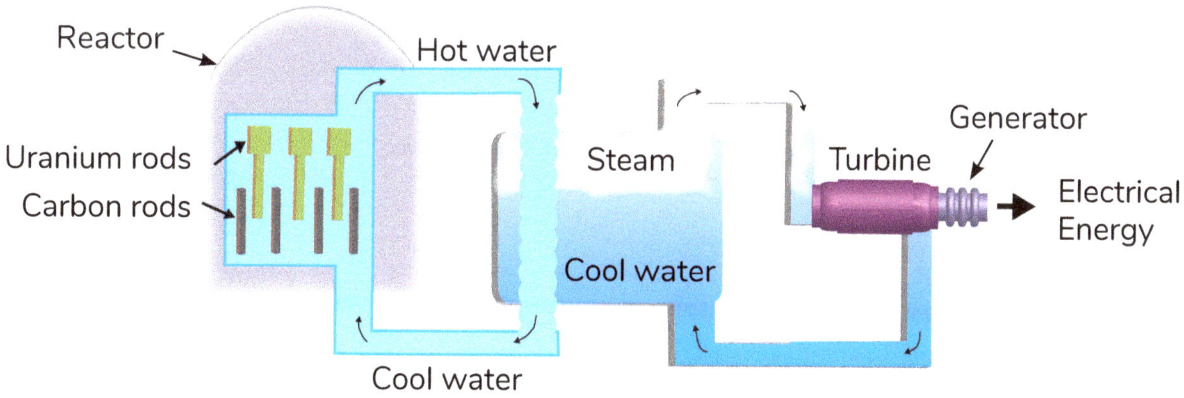

16.5 Summary

- **Natural resources** include **fossil fuels** and **renewable fuels**.
- **Oil**, **natural gas**, and **coal** are types of **fossil fuels**.
- **Sun**, **wind**, and **water** are types of **renewable fuels**.
- Wind and water systems use **turbines** and **generators** to convert water and air flow into electricity.
- **Solar panels** are made of **photovoltaic cells** and convert sunlight directly to electricity.

16.6 Vocabulary

Basic Words

deplete (dih-PLEET)

distill (dih-STIHL)

distillation (dih-stihl-AY-shuhn)

fission (FIH-shuhn)

fossil fuel (FAH-suhl FYOOL)

fraction (FRAAK-shuhn)

krypton (KRIHP-tahn)

refine (rih-FIYN)

resource (REE-sawrss)

turbine (TUHR-buhn)

Challenge Words

barium (BAA-ree-uhm)

generator (JEH-nuh-ray-tuhr)

hydroelectric (hiy-droh-ih-LEHK-trihk)

natural (NAA-chuh-ruhl)

nuclear (NOO-klee-uhr)

photovoltaic (foh-toh-vahl-TAY-ihk)

renewable (rih-NOO-uh-buhl)

uranium (yoo-RAY-nee-uhm)

CHAPTER 17

Threats and OPPORTUNITIES

17.1 Introduction

Earth is a mostly self-contained global ecosystem. Energy in the form of light and heat enters and leaves our global ecosystem, but an appreciable amount of matter neither enters nor leaves our planet. A system that is fully self-contained is called a **closed system**, and although our planet is not a totally closed system, it can behave geologically, biologically, and energetically as a closed system.

Because our global ecosystem behaves like a closed system, life can exist and thrive only when the various matter and energy cycles are in balance. Because we cannot always predict how our global ecosystem will respond to changes in how we manage our use of matter and energy, life on Earth is faced with both threats and opportunities.

17.2 Trash

One problem with modern life is the generation of trash. When you go to the grocery store and buy a basket of strawberries, you are not only taking home the strawberries but, depending on where you shop, you might also bring home a plastic basket and a plastic bag. You eat the strawberries, but often you throw away the plastic basket and plastic bag. The plastic basket and plastic bag become trash that eventually ends up in a landfill.

It is estimated that American households produce over 300 million tons of trash each year. On average, a person produces about 5 pounds (2.3 kilograms) of trash a day. Most of this trash goes into landfills. Unfortunately, landfills are not the best place for trash. Trash accumulates in landfills and often does not readily decompose. Landfills become full and then new landfills are created to house more and more trash. This accumulation of massive amounts of trash creates a threat to long-term life on Earth because chemicals that are thrown out can seep into the groundwater and materials that decompose can give off gases that pollute the atmosphere.

One way to manage the accumulation of trash is to **recycle** it. Recycling uses the idea that ultimately all matter is made of atoms. People convert usable matter, like carbon and hydrogen atoms, into forms that don't easily break down in nature, like plastic and Styrofoam. However, with the use of chemistry, people can find ways to convert plastic and Styrofoam back into carbon and hydrogen. Recycling is a way to take trash and convert it back into usable matter.

Recycling is complicated and it takes a great deal of science and engineering to figure out how to convert different materials from one form to another. Plastic, glass, and paper can all be recycled, but each one requires a different process. Also, because there are different types of plastic, different types of glass, and different types of paper, not all plastic, glass, or paper can be recycled in the same way. The managing and recycling of trash creates a huge opportunity for new research and technology and the promise of new jobs. Most importantly, being able to recycle unusable materials into usable materials helps bring balance back to crucial matter and energy cycles.

17.3 Air and Water Pollution

Another problem with modern life is the generation of air and water **pollution**. Just like trash, pollution in air and water comes from the conversion of usable matter, such as oxygen and nitrogen, into matter that can harm the environment.

When something is burned, there is a chemical reaction. Many modern processes require the burning of fuels such as coal and gas. When coal and gas are burned, carbon dioxide and water are produced. However, nitrogen oxides and sulfur compounds are also produced as **by-products**.

In chemistry, a by-product is a molecule or set of molecules that are produced in addition to the **primary products**. For example, when coal and gas fuels are burned, carbon dioxide and water are the primary products. But gas and coal contain small amounts of sulfur, and the surrounding air contains nitrogen. When sulfur is burned, it combines with oxygen in the air and sulfur dioxide is produced. When nitrogen from the air is burned, it combines with oxygen to produce nitrogen dioxide. Sulfur dioxide and nitrogen dioxide are by-products of burning coal and gas fuels.

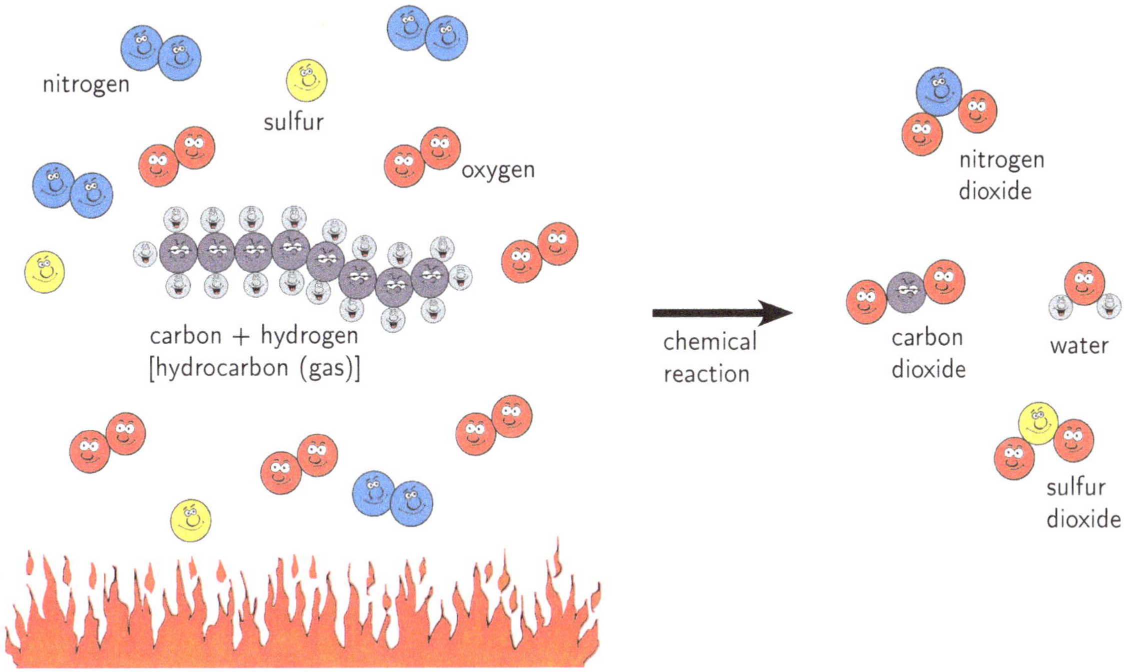

When nitrogen dioxide and sulfur dioxide, the by-products of burning fossil fuels, are released into the atmosphere, they can cause air pollution and **acid rain**.

Acid rain occurs when nitrogen and sulfur by-products in the air interact with sunlight and water to produce nitric acid and sulfuric acid. These acids mix with the water in the atmosphere, and when it rains, the acid in the raindrops falls on cities and farmland. Over the years acid rain can destroy buildings, forests, lakes, agricultural crops, art sculptures, and monuments.

In addition, when sunlight hits by-products from the burning of fossil fuel, it can trigger a set of chemical reactions that produces **ozone**. An ozone molecule is made of three oxygen atoms. Ozone is necessary for the lower stratosphere, but at ground level it creates a stinging gas that can cause damage to plants and animals. The brown stuff, or **smog**, that hangs over many large cities is due to air pollution, including ozone.

Pollution in our oceans, rivers, lakes, and underground aquifers is another problem caused by modern life. Many industrial processes produce different by-products that are sometimes released into the Earth's waterways. Many people around the world do not have access to clean water because sewage and industrial wastes have entered the water system.

Removing and preventing air and water pollution is a major issue that represents both a threat to life on Earth and an opportunity for science and engineering research and development. Humans will continue to make, build, and create new products that will alter our environment. However, by discovering new technologies, all of the materials, chemicals, and pollutants humans make can be converted back to atoms and molecules that are safe for the environment, helping to maintain the balance of important matter and energy cycles.

17.4 The Greenhouse Effect

Greenhouses are enclosed structures made of glass or plastic. The glass or plastic lets sunlight enter and then traps some of the heat from the Sun, holding it inside the greenhouse. Trapping the heat keeps a greenhouse warm inside even during winter, allowing plants to grow year round. Our planet's atmosphere traps energy from the Sun in a similar way, and this is called the **greenhouse effect**. The greenhouse effect occurs because energy from the Sun enters Earth's atmosphere, but not all of it escapes back into space. Some gases act like a greenhouse by absorbing heat and keeping Earth warm. These gases include carbon dioxide, water vapor, and methane, among others.

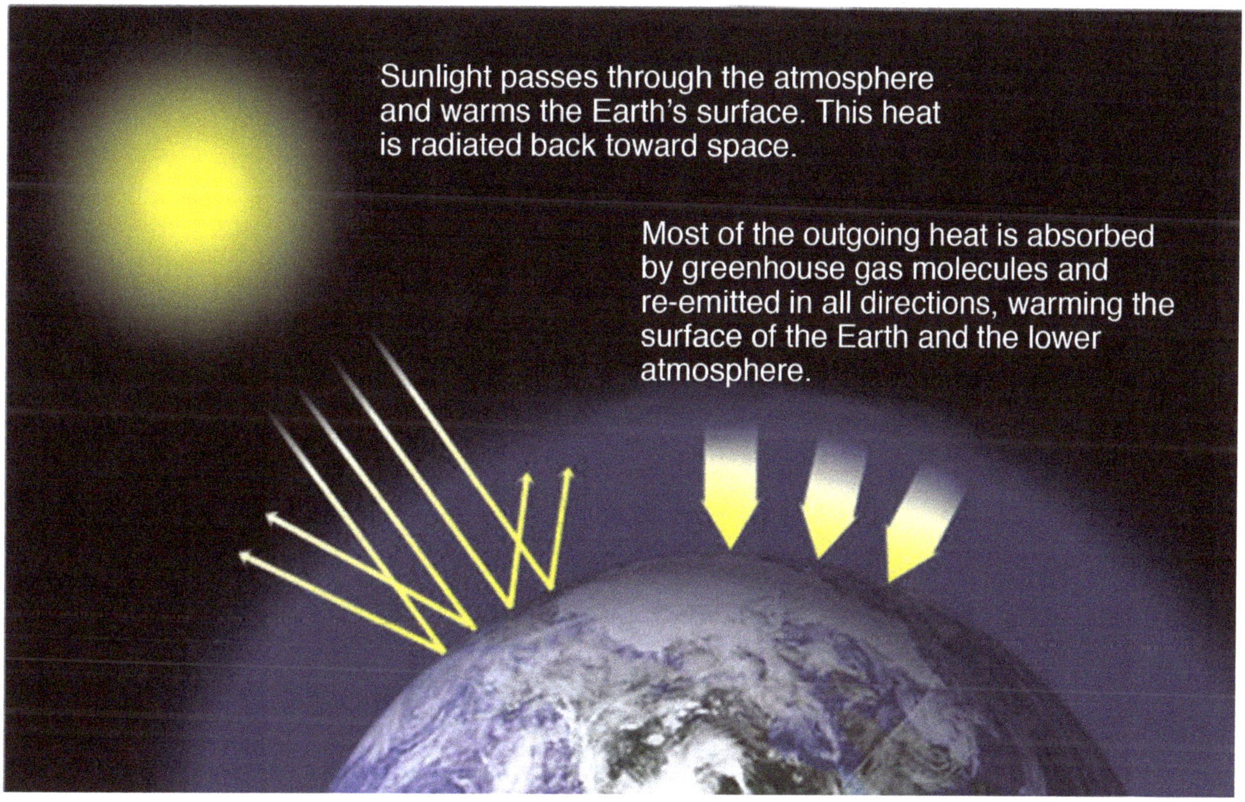

During the day, sunlight warms the planet. At night the surface of the Earth cools and releases heat back into the atmosphere. Some of this heat does not escape into space but is trapped by **greenhouse gases**. If this process remains balanced, the Earth will stay at about the same yearly average temperature. If the greenhouse effect is too strong, it won't allow as much heat to escape into space, and then Earth begins to get warmer.

Every day scientists collect a lot of data about air and ocean temperatures all over the world. The data show that currently there is a rise in the average temperature of the Earth's atmosphere and oceans. Some indications of a warming of Earth are record high temperatures at different locations, the melting of glaciers, severe flooding, droughts, and fires. Although the warming trend is clear, the causes of the warming are under debate.

Measurements show that there is an increase of greenhouse gases in the atmosphere. Scientists believe that human activity is causing Earth's warming trend and think greenhouse gases are coming mostly from the burning of fossil fuels. Because fossil fuels like oil, coal, and natural gas are used for many things, including generating electricity, heating buildings, cooking, and running machinery, large amounts of carbon dioxide are released into the atmosphere.

Data collected show that **deforestation** (the cutting down or burning of trees) in the Amazon rainforest leads to localized temperature increases, increased CO2 emissions, and lack of rainfall due to disruption of the water cycle. Scientists predict that if deforestation in the Amazon continues, it could negatively impact global temperatures and water cycles.

Other scientists think that Earth is in a natural warming cycle, while still others think that there are a variety of factors involved. Scientists often have different theories about the same thing, and the warming of Earth is one area of study in which there are different ideas.

17.5 Summary

- Earth operates mostly as a self-contained **closed ecosystem,** but energy enters and leaves.
- Problems caused by modern life are the production of trash, air and water **pollution**, and the increase in the **greenhouse effect.**
- Overcoming the threats caused by trash, air and water pollution, and the greenhouse effect creates opportunities for new scientific and technology developments.

17.6 Vocabulary

Basic Words

acid (AA-suhd)

effect (ih-FEHKT)

greenhouse (GREEN-howss)

ozone (OH-zohn)

smog (SMAHG)

Challenge Words

by-product (BIY-prah-duhkt)

deforestation (dee-faw-ruh-STAY-shuhn)

ecosystem (EE-koh sih-stuhm)

pollution (puh-LOO-shuhn)

primary (PRIY-mehr-ee)

recycle (ree-SIY-kuhl)

CHAPTER 18

Exploring the UNIVERSE

18.1 Introduction

Have you ever wondered what it would be like to take a rocket to the Moon? Do you think people can live on Mars? Do you think humans could live on Titan, Saturn's largest moon with its dense atmosphere of nitrogen and methane and the possibility of liquid water?

In the past, exploring the universe happened only from Earth's surface, using telescopes and mathematics. But even before man first stepped on the Moon in 1969, scientists and explorers alike have been thinking about how to make space travel to distant planets and galaxies an everyday reality.

What would it take to travel through the solar system and visit distant planets? How far is it to the planets? How long would it take to get to them, and how fast would you need to travel? What type of spaceship would you need, and how would you carry all the necessary supplies with you, such as food, water, equipment, and fuel? In this chapter, we'll take a look at what it might take to explore the universe with space travel.

18.2 How Far?

How far is it to the Moon? Can we travel to Mars? How about Venus, Jupiter, or the planets around a distant star — can we travel that far?

Think about your neighborhood. How big is your neighborhood? Can you travel on your bike to the corner store or do you need to take a car? What about traveling across your city, state, province, or country? What type of motor vehicle, airplane, or train would you need? How long would it take?

In planning travel to a distant place, the first thing to figure out is how far the place is from your home. If you want to go down the block to visit a friend, you might choose to walk. If your neighborhood isn't very big, your friend's house isn't very far away, and you know that walking from your house to your friend's house is something you could easily do in a relatively short amount of time, then walking is a good choice.

However, if you want to travel to a friend's house in another part of the city, you might take your bike or even talk your parents into giving you a ride in the car. You'd choose a different way to travel because you know that the bigger the distance between you and your friend's house, the farther it is to travel and the longer it will take you to get there. If you want to visit a friend in another town, walking for days isn't a good choice when a car could get you there in a few hours.

When it comes to exploring the universe, a major question is, "Just how big is the universe?" We know how big Earth is and we have a good idea how big our solar system is, but exactly how big is our galaxy? And how big are all the other galaxies that fill the space we call the universe? How far away are they?

The Moon is the closest celestial body to Earth, and yet the distance from Earth to the Moon is an average of 384,400 km (238,900 miles). If you could travel to the Moon using a jet plane going 1600 km per hour (1000 miles/hr), it would take 10 days to reach the Moon.

Using **rockets** that go much faster than a jet, **astronauts** can reach the Moon in only a few days. In 1969 after four days of travel, the three astronauts on **NASA's Apollo 11 mission** were the first to land on the Moon. The Apollo 11 astronauts reached a speed of about 39,000 km per hour (24,000 miles per hour) in order to escape the pull of Earth's gravity. After leaving Earth's gravity, the speed of the flight slowed but was still much faster than a jet could travel.

It is easy to imagine traveling to the Moon because the Moon orbits Earth and is relatively close. But what if we want to travel to Mars? Mars and Earth both orbit the Sun, but the distance from Earth to Mars is constantly changing. When Earth and Mars are on the same side of the Sun and closest to each other, the distance between them is about 55 million km (34 million mi.).

It currently takes spacecraft about seven months to reach Mars at the time when the orbit of Mars brings it close to Earth. However, if Earth and Mars are on opposite sides of the Sun, not only is the distance between the planets much farther at 401 million km (249 million mi.), but this distance is measured straight through the Sun.

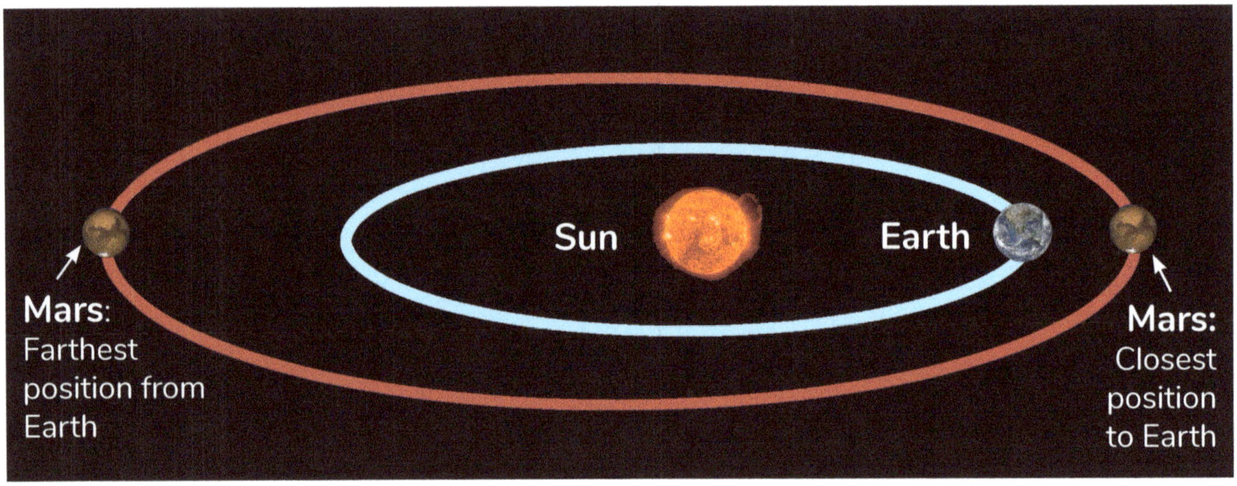

Traveling to any celestial body other than the Moon gets more complicated very quickly. Not only are the distances much longer, but the distances and positions of the celestial bodies are constantly changing. To get to another planet, calculating the route, how the planet is moving, what type of spaceship and fuel to use, how long the trip will take, and what the astronauts will need to live requires many scientists and engineers using physics, chemistry, geology, astronomy, biology, advanced technologies, and mathematics.

18.3 Rockets

As mentioned before, on July 16, 1969 **NASA (National Aeronautics and Space Administration)** launched three astronauts into space. For the first time in history, humans walked on the Moon during the Apollo 11 mission.

To get to the Moon, the astronauts used a 111 meters tall (363 feet) rocket called the **Saturn V** (Saturn Five). The Saturn V was a rocket specifically designed to carry people and heavy equipment. It stood 36 stories tall and weighed as much as 400 elephants. It was the most powerful rocket that had ever been launched successfully.

To get to the Moon, this rocket needed several main parts. First, it needed a **frame** to give it structure. A rocket frame is like the **fuselage** (main body) of an airplane. It has to be as light as possible and also must be very strong. Most rocket frames are made of **titanium** and **aluminum**, which are strong, lightweight metals. A rocket frame is cylindrical in shape like an airplane fuselage.

A rocket going to the Moon needs to carry a **payload**. For the Apollo 11 space flight, the astronauts and their spacecraft were the payload. The payload is located at the top of a rocket.

The Saturn V rocket

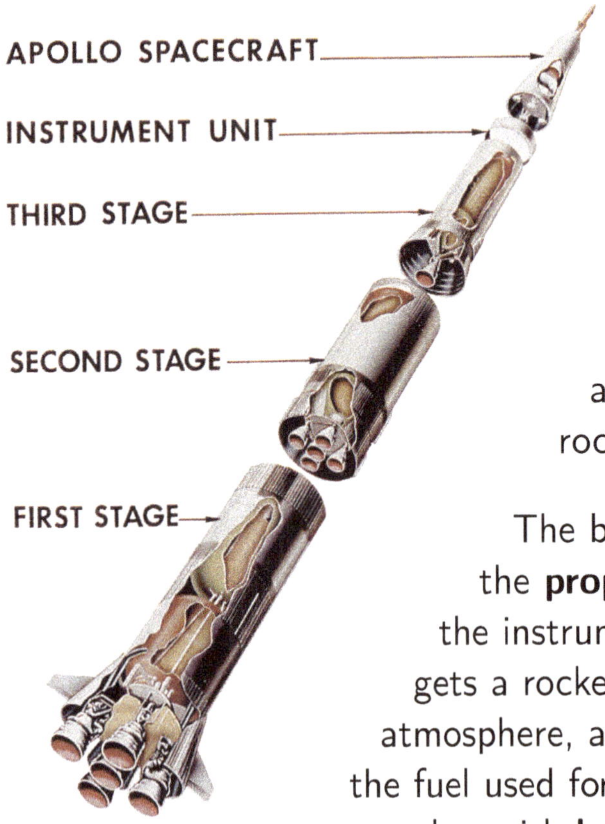

Behind the payload is the **instrument unit** which contains the **guidance system**. The guidance system is the equipment that guides the rocket and keeps it on the correct path. The guidance system includes technology such as sensors, computers, communication devices, and the equipment to steer the rocket.

The bulk of a rocket is made up of the **propulsion system** which is behind the instrument unit. A propulsion system gets a rocket off the ground, through Earth's atmosphere, and into space. For the Saturn V, the fuel used for propulsion was **liquid oxygen** together with **kerosene** and **liquid hydrogen**.

Propulsion systems for space rockets are organized into **stages**, or sections, with each stage having its own engines. The Saturn V had three stages. The first of the three stages launched the rocket from the rocket platform into the atmosphere and carried 770,000 liters (203,400 gallons) of kerosene and 1.2 million liters (318,000 gallons) of liquid oxygen. Once the height and speed needed were attained, the first stage separated from the rocket and fell back to Earth.

The second stage of the Saturn V carried 984,000 liters (260,000 gallons) of liquid hydrogen and 303,000 liters (80,000 gallons) of liquid oxygen. The second stage propelled the rocket farther into the atmosphere and was released when its function was complete.

When the second stage released, the third stage engines fired. This stage carried 252,750 liters (66,770 gallons) of liquid hydrogen and 73,280 liters (19,359 gallons) of liquid oxygen. After the third stage fired, the rocket began to orbit around the Earth. Then, at the correct time, the third stage engine was fired again and sent the spacecraft out of its orbit around the Earth and toward the Moon.

Saturn V rockets were used by NASA from 1967-1973 and then retired. Since then, technology has advanced and been refined, but rockets today still work in the same basic ways, with stages and some of the same types of fuels. Rockets today often use **solid fuels** in addition to liquids.

From 1981-2011 NASA used five **space shuttles** that were a combination of rocket, aircraft, and glider. A space shuttle looked like a big airplane with a huge rocket pack attached. The part that looked like an airplane was called the **orbiter**. Attached to the orbiter were the fuel tank and two rockets that launched the orbiter using a chemical propulsion system of both liquid and solid fuels. The orbiter's wings were used for gliding to the surface of the Earth and wheels were used for landing.

The space shuttles carried astronauts and satellites into space. They transported equipment and materials to build and repair the **International Space Station (ISS)** and restock it with food, and they carried astronauts back and forth between Earth and the ISS. More recently, spaceships from commercial companies bring most of the equipment and supplies to the ISS and transport crews back and forth.

18.4 The Future of Space Travel

What does the future look like for space travel? Do you think we will one day have people flying back and forth from all the planets in the solar system? What about intergalactic travel (travel to other galaxies)? Will we one day be able to travel beyond the Milky Way to explore some of the many billions of other galaxies in the universe?

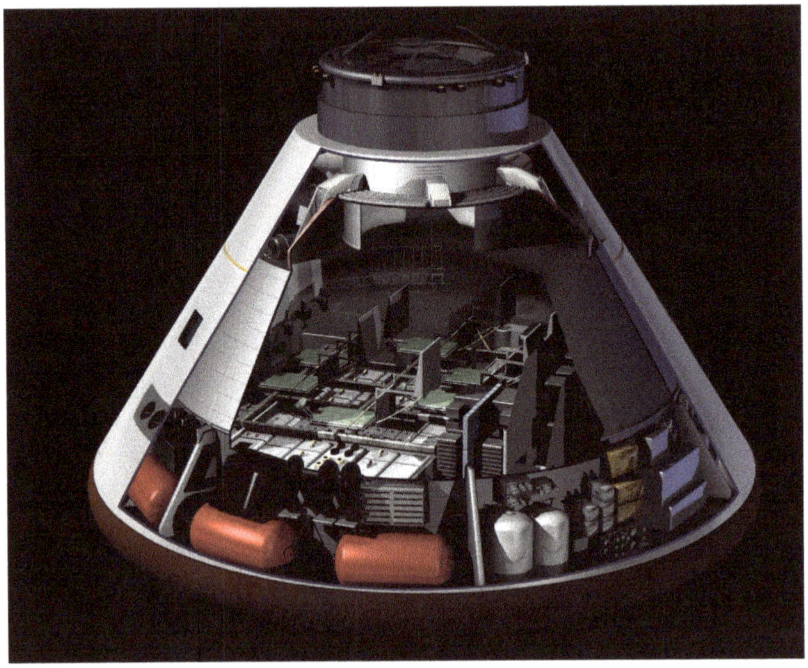

NASA's Orion spacecraft crew module designed to take a crew of four on flights as far away as Mars

The future of space travel depends largely on how quickly we can move through space, how much cargo can be carried, and finding ways to replenish supplies. With current technology, even traveling to a nearby planet in our solar system would take a long time because the planets are so far away.

Traveling quickly through space means finding a fuel that is efficient and powerful and that doesn't take up too much space. Liquid fuels, such as liquid oxygen, liquid hydrogen, and kerosene, require specialized tanks that are heavy and take up a significant amount of space. Solid fuels are more compact and take up less space. Because neither liquid nor solid fuels can be easily replenished in space, the problem of carrying enough fuel adds to the difficulties of planning long-distance space travel.

Currently, different types of propulsion methods are being explored. Some utilize chemical propulsion — fuels that are liquid, solid, gels, or a combination. Others may use electric and magnetic fields, solar energy, lasers, or nuclear energy. Some scientists are exploring a deeper understanding of physics concepts such as space-time and gravitation to try to find a totally new form of propulsion.

It is possible that there is an entirely new way to travel through space that we have not yet discovered. Imagine being able to fold space or teleport through time like we see in the movies. It may seem crazy to us today, but as new discoveries in science make new technologies possible, currently unimaginable ways to travel may become a reality. At one time, putting a man on the Moon sounded crazy.

This is a very exciting time in the history of astronomy. Humanity is getting ready to explore the universe in person. Many new ideas are being developed for traveling and living off Earth. For example, it may be possible to use inflatable modules to build new space stations to orbit Earth. These inflatable units could be folded up when they are launched, making a smaller package to lift into space. They would also be lighter and stronger than modules made of metal. When the modules reach their destination, they would be inflated and hooked together.

Technology is advancing rapidly. Habitats that people can live in on other planets are being tested. Bases on the Moon and the mining of asteroids are being planned. Scientists are exploring 3D printing technology to see how it can be used to build habitats and make tools and food in space. Spaceports are being built in countries around the world, getting ready for the time when people travel into space on a regular basis.

This is a great time to be a scientist. Maybe you will be the one to come up with a new discovery that advances space travel.

18.5 Summary

- Because the distances to be traveled between planets are so large, designing new types of spacecraft is important.
- Rocket design includes a lightweight **frame**, a **payload**, a **guidance system**, and a **propulsion system**.
- Scientists are researching new ways of traveling through space, and new technologies are being developed for living off Earth.

18.6 Vocabulary

Basic Words

guidance (GIY-duhns)

NASA (NAA-suh)

payload (PAY-lohd)

rocket (RAH-kuht)

Saturn V (SAA-tuhrn FIYV)

system (SIH-stuhm)

Challenge Words

administration (uhd-mih-nuh-STRAY-shuhn)

aeronautics (ehr-uh-NAW-tihks)

aluminum (uh-LOO-muh-nuhm)

Apollo 11 (uh-PAH-loh ih-LEH-vuhn)

astronaut (AA-struh-nawt)

fuselage (FYOO-suh-lahzh)

instrument (IHN-struh-muhnt)

propulsion (pruh-PUHL-shuhn)

seismic (SIYZ-mihk)

titanium (tiy-TAY-nee-uhm)

CHAPTER 19

How Big Is the UNIVERSE?

19.1 Introduction

As we saw in the last chapter, one of the biggest challenges to space travel is the fact that the distance between Earth and each of the other planets in our solar system is so vast. With our current technology it would take a long time for astronauts to travel to another planet. Spacecraft would need to carry enough fuel and supplies for astronauts to survive a long trip.

Long ago, people believed that the universe was just a little bit larger than Earth, but today we know that the universe is so large that it is hard to imagine its size. With modern telescopes, space probes, and other technologies, we can measure some of the universe. However, astronomers still don't know the exact size of the universe. Also, as advances are made in technology, scientists keep discovering more celestial bodies beyond those already known.

Attempting to measure the universe is difficult because there are such huge distances involved. We can't take a tape measure and hop from Earth to the end of the universe to make a measurement. Scientists and astronomers must come up with other ways to measure distances in space.

19.2 How Do We Measure "Big"?

You have probably used both rulers and tape measures to find out how big something is. But how do we measure distances that are much longer than a ruler or tape measure can reach? How can we measure the distance to a faraway galaxy?

Measuring the distance from Earth to the Moon, the planets, and the Sun is not an easy task. The first astronomer to find a way to measure the distance from Earth to the Moon was **Aristarchus of Samos**, a Greek astronomer and mathematician who lived from 310-230 B.C.E. Aristarchus observed how the Moon passes through Earth's shadow during a lunar eclipse. With geometry, Aristarchus used this information to calculate the approximate distance between the Earth and the Moon. He also used geometry to calculate an approximate distance from Earth to the Sun.

Aristarchus of Samos

310 230 B.C.E.

Today, astronomers use a phenomenon called **parallax** to make accurate measurements of the distance to stars that are closer to Earth. Parallax is the effect that makes a nearby object appear to shift position when it is viewed from different locations.

You can see the parallax effect by using your own thumb. Stretch out your arm and hold your thumb up in line with your nose. Close your right eye and line up your thumb with an object in the background, for example, a window frame, door frame, or tree. Keeping your thumb in the same position, close your left eye and open your right eye. Did your thumb appear to move? Now repeat this experiment while first holding your thumb close to your face and then holding it as far as you can from your face. Do you notice that your thumb appears to move more when you hold it close to your face than when you hold it farther away?

The "motion" of your thumb is called parallax. Your thumb didn't actually move, but because each of your eyes is in a different position, the angle at which you view your thumb is different for each eye. Each of your eyes sees your thumb in a different position relative to the background.

Astronomers use the same principle of parallax to measure the distance to nearby stars. Because of the parallax effect, stars that are closer to us appear to shift position in the sky as Earth moves in its orbit around the Sun. Stars that are much farther away don't appear to change position.

To find out the distance to a nearby star, astronomers begin by measuring the position of the nearby star relative to another star that is very far away. Six months later when the Earth's orbit has taken it to the other side of the Sun, the positions of the two stars are measured again. The nearby star will appear to have moved and the very far away star will be in the same position. Because astronomers know how far Earth traveled in six

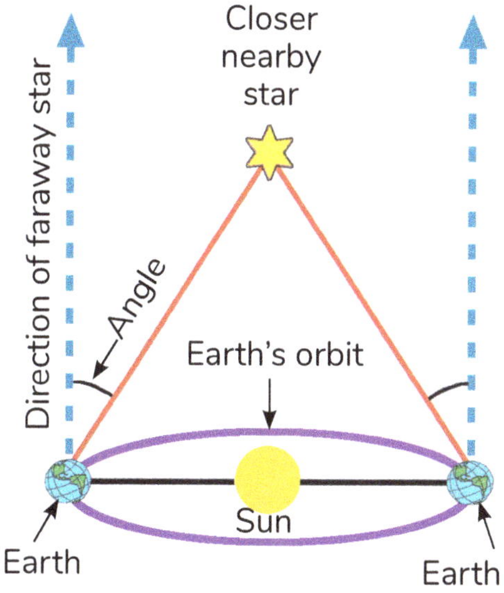

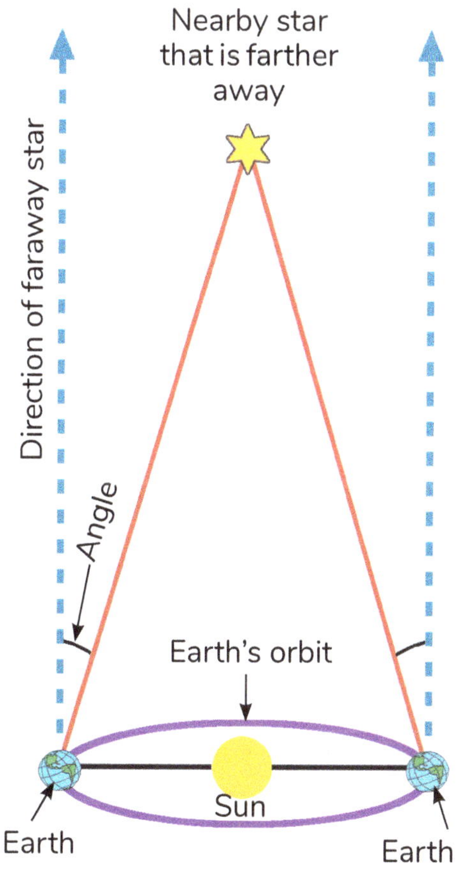

Using parallax to measure distance to a star

months, they can use this number along with the relative positions of the nearby star and the very faraway star to do mathematical calculations that will tell them how far away the nearby star is. In the illustration, the angles are shown as being large, but in astronomy these angles are very, very small because the stars are so very far away.

Measuring the distance of stars by parallax is a very slow method, but advanced technology can make it faster. In 1989 the **European Space Agency (ESA)** began using a satellite called **Hipparcos** to measure parallaxes much more quickly. In only four years Hipparcos accurately mapped over 100,000 stars.

The **GAIA** satellite was launched In 2013 by ESA. GAIA has mapped about two billion stars along with their positions, distances, brightness, motions, what they are made of, and other information.

Artist's impression of the GAIA satellite

Another way to measure the distance of faraway celestial objects is to use **radio telescopes** such as the **Very Large Array** in New Mexico. Celestial bodies give off **radio waves**, which are low frequency electromagnetic waves that can be detected by radio telescopes. By using computers and complicated mathematics, astronomers can calculate how far the radio waves have travelled from a celestial body and can learn many other interesting facts about the object.

Astronomers use orbiting telescopes and space probes to carry instruments that can detect other kinds of electromagnetic waves, such as **infrared**, **gamma rays**, and **X-rays**. When telescopes on Earth and instruments on spacecraft are close enough to a celestial body, radio waves from the instruments can be bounced off the object. The radio waves that are reflected back to the instruments contain a lot of data about the object.

Although we can measure distances accurately using parallax, electromagnetic waves, space probes, and other technology, we still don't know how big the universe is. It seems that as technology advances and we are able to detect objects that are farther and farther away, we keep finding that the universe is bigger than we thought.

Faraway galaxies

19.3 Does the Universe End?

If the universe ends, is there something beyond the universe? Astronomers have found evidence that supports the theory that the universe is expanding. But if the universe is expanding, it needs to have a place to expand into. What is that place? More space? Will the universe continue to expand? Or is there a limit to how far the universe can expand? These are among many questions that astronomers and astrophysicists ask while studying the universe.

One idea is that the universe doesn't end but is **infinite**. Infinite means that there is no limit or ending. Infinity is a difficult concept to grasp because we experience everything, including space, as having limits.

For example, your house has a limit you can measure. You can walk around the outside of your house and see where your house begins, where it ends, and what lies beyond it. The same is true with your city. Your city has a city limit, or boundary, and anything outside your city is either part of a rural area or part of another city or town. Your state or province and country have a limit. Our planet, solar system, and galaxy also have limits. But we do not know if there is a limit to the universe.

It makes sense that as we keep moving farther and farther from Earth, we would discover that space itself and the universe have a limit. But so far there is no evidence to suggest that the universe ends. In other words, we have not yet found the limit of the universe even by using our most advanced technologies. This doesn't mean that there isn't a limit to the universe, it just means that if a limit exists, we are not able to find it with the technology we now have.

19.4 Summary

- The distance from Earth to individual stars can be measured.
- Electromagnetic waves coming from celestial bodies can be gathered by **radio telescopes** and other instruments to provide information.
- It is unknown whether there is an end to the universe.

19.5 Vocabulary

Basic Words

array (uh-RAY)

ESA (EE-suh)

GAIA (GIY-uh)

gamma ray (GAA-muh ray)

Challenge Words

Aristarchus of Samos (aa-ruh-STAHR-kuhs uhv SAY-mahs)

Hipparcos (hih-PAHR-kohs)

infinite (IHN-fuh-nuht)

infrared (ihn-fruh-REHD)

parallax (PEHR-uh-laaks)

radio telescope (RAY-dee-oh TEH-luh-skohp)

CHAPTER 20

Are We ALONE?

20.1 Introduction

While looking at the many stars that light up the night sky, have you ever wondered if we are alone? Is Earth the only planet in the universe that supports life? Is carbon-based life, the life that we know, the only kind of life possible? Could there be a planet in outer space with intelligent life?

These are fun questions to ask and they make great plots for movies and works of fiction, but they have been difficult for scientists to answer. The only life we have direct evidence for is life on Earth. To date, scientists have not found any evidence that life exists on other planets.

However, with so many stars and so many galaxies decorating the universe, the possibility that life exists on other planets intrigues not only movie directors and novelists, but also scientists and astronauts as well. If life exists on other planets, can we find it?

There are two ways to approach the search for life in the universe. The first is to look for planets in other solar systems that have the right conditions to support carbon-based life, like ours. The other way is to look for signs of life, and particularly intelligent life, that may or may not come from a carbon-based organism.

20.2 Looking for Life in Our Solar System

The only type of life we know about is life that is made primarily of carbon. All life on Earth contains DNA and proteins and other molecules made primarily of carbon, hydrogen, and oxygen. Life as we know it is carbon-based. Because carbon-based life needs water and oxygen to survive, astronomers look for evidence of liquid water and the right kind of atmosphere on other planets as a way to determine if life as we know it is possible.

Both the Moon and the planet Mercury lack liquid water and an atmosphere. Neither has a protective magnetic field, so they are easily affected by bombardments of solar wind, asteroids, and cosmic rays.

The Moon

Mercury

Venus

Venus has no liquid water and has a very thick and very hot atmosphere made mostly of carbon dioxide. The clouds that cover Venus are full of corrosive sulfuric acid. Because of these conditions Venus isn't suitable for carbon-based life.

Jupiter, Saturn, Neptune, and Uranus are gas planets. Because they lack a rocky surface and don't have liquid water or the proper atmosphere, none of these planets would be suitable for life as we know it. However, some of the moons that orbit these planets may be candidates for supporting life. Jupiter has several moons that look promising. For example, **Europa**, a moon of Jupiter, is thought to have an ice-covered ocean of water with a rocky sea floor.

Europa and a closer view of its icy surface

One way to determine if life is possible on Europa is to look for life in extreme environments on Earth. Organisms that live in extreme conditions are called **extremophiles**. Extreme conditions include very high temperatures and pressures, very low temperatures, and a lack of oxygen. Extremophiles are often single celled bacteria and archaea. If life can exist on Earth in extreme conditions that may be similar to those on Europa, then it might be possible for life to exist in Europa's oceans.

Mars

Mars is the planet in our solar system that scientists think may at one time have had the conditions needed for some type of life to exist. Mars is a rock planet with a solid surface like Earth. The Martian atmosphere is thinner than Earth's and is made mostly of carbon dioxide with only a very small amount of oxygen. Mars is smaller than Earth with a longer year and colder temperatures. Its axis is tilted, creating seasons. Mars has no liquid water but does have water in the form of ice. In the past Mars had surface water as can be seen by water's effects on the landscape.

Because Mars is relatively close to Earth and because we are so curious about Mars, scientists have launched probes, landers, and rovers to explore the planet, and more Mars missions are being planned. NASA states that it has four goals in exploring Mars: to find out whether life ever existed on Mars, to learn more about the martian climate, to learn about the geology, and to prepare for humans to go to Mars.

A **rover** is a spacecraft that lands on a moon or planet and explores it by traveling on its surface and taking photographs, sampling rocks and soil, and studying environments. NASA has sent several rovers to Mars.

In 2011 NASA launched a rover called **Curiosity**, which is more than just a camera on wheels. Curiosity has a whole mobile laboratory on board. Curiosity can take photos and perform experiments using lasers, drills, and specialized equipment and is looking for evidence that past conditions on Mars could have supported life. Curiosity has been able to confirm that flowing water once existed on Mars, but so far no signs of life have been detected.

Perseverance rover being lowered to Mars from a spacecraft
Illustration

NASA's **Perseverance** rover landed on Mars in February 2021. Perseverance is studying the rocks and landscape of Mars and looking for signs that life existed in the distant past. It is also collecting and packaging rock and soil samples that will later be picked up and returned to Earth by other spacecraft. Scientists are using the Perseverance mission to learn more about the conditions on Mars to help with the preparations for humans to land and stay there.

20.3 Looking for Life in Our Galaxy

As far as we know now, it appears that Earth is home to the only intelligent life in our solar system. Therefore, if we are going to find life on other planets, it seems reasonable to extend the search beyond our solar system and look at planets in other solar systems in our galaxy. Planets that are outside our solar system are called **exoplanets**. Many exoplanets have been discovered with space telescopes.

Size comparison of Earth (far right) to some exoplanets found by the Kepler Space Telescope

Because exoplanets are so far away, we can't send a rover to them to perform experiments and test for liquid water. But we can use space telescopes to look for planets that appear to be the right distance from their sun to have the possibility of supporting life as we know it.

Why do planets need to be the right distance from their sun? If a planet orbits too close to its sun, the surface of the planet is too hot to support life. If a planet orbits too far from its sun, the surface of the planet is too cold to support life. Only planets that are just the right distance away from their sun are able to support life as we know it. These planets are said to be in the **habitable zone**. Although astronomers now think that most stars have planets orbiting them, only some of those planets are the right distance from their sun that life as we know it might exist.

Illustration showing our solar system's habitable zone compared to that of a two planet/two star solar system.
The habitable zones are in green.

Earth-like planets outside our solar system are hard to see even with advanced telescopes. An Earth-like planet will reflect light from its sun, but it will be a very weak reflection, which means our telescopes won't easily see it. Also, when an Earth-like planet is close enough to its parent sun that it might support life, the weak sunlight it reflects will be overcome by the bright light of the parent sun. So how can we look for Earth-like planets that are just the right distance from their sun if they are so difficult to see?

Illustration of exoplanets transiting their sun

One way to look for planets is to observe **transits**. A transit happens when a planet crosses in front of its sun. Transits by planets will produce a small change in the star's brightness, and this change can be detected. Earth-like planets have been found by observing transits.

Using space telescopes, astronomers have discovered thousands of exoplanets in our galaxy and are also expecting to find exoplanets in other galaxies as technology advances. Data from the space telescopes is analyzed to discover how far a planet is from its sun, what gases are in its atmosphere, and what the planet is made of. Some of the exoplanets are in habitable zones and are of particular interest.

20.4 Looking for Faraway Life

Exploring far distances in space and finding planets that may be able to support life is challenging. Although there are billions of galaxies, stars, and solar systems out there, the size of the universe makes it hard to find these planets.

The search for life on other planets brings up many questions. What is life? What is a sign of life? What would be evidence of life as we know it? What would be evidence of life as we don't know it? Will we know life when we see it if it is very different from ours?

Since we can't yet travel to any exoplanets, one way to look for life on other planets is to look for signs of intelligence. Every day, humans send out electromagnetic signals that indicate the existence of intelligent life on Earth. Recall that the music you hear on your radio is made of radio waves traveling at the speed of light. Radio waves that transmit music or speech have a certain pattern that can be distinguished from noise, and these radio waves travel out into space from Earth. Because radio waves have long wavelengths, they are less likely to be scattered by small particles in interstellar space.

A radio telescope array

Looking for radio waves that have a particular pattern is one way to look for intelligent life. Astronomers use groups of telescopes called **arrays** to gather radio waves coming from space in hopes of finding signals that confirm intelligent life. A few intentional radio transmissions have been sent out from Earth in hopes of receiving a signal back, but some think that it would take far too long for the radio waves to reach an exoplanet and then be returned by intelligent life for this method to be useful in gathering data.

Artificial light on a planet is also looked for since it might indicate the presence of cities. A planet's atmosphere can be checked for signs of pollution or manufactured chemicals that could indicate advanced civilizations. Other kinds of gases could be produced by life forms.

Finding and studying exoplanets and looking for life are rapidly advancing areas of astronomy with many exciting new discoveries waiting to be made.

20.5 Summary

- In our solar system, Mars and some of the moons of Jupiter are possible candidates for having some form of life now or in the past.
- Astronomers look outside our solar system for Earth-like planets that are the right distance from their sun to possibly have life.
- Astronomers can find **exoplanets** by observing them as they are **transiting** their sun.
- Astronomers look for intelligent life with space telescopes and Earth-based radio telescopes.

20.6 Vocabulary

Basic Words

array (uh-RAY)

rover (ROH-vuhr)

transit (TRAAN-suht)

Challenge Words

Curiosity (kyoor-ee-AH-suh-tee)

Europa (yoo-ROH-puh)

exoplanet (EHK-so-plaa-nuht)

extremophile (ihk-STREE-muh-fiyl)

habitable (HAA-buh-tuh-buhl)

Perseverance (puhr-suh-VIHR-ehns)

CHAPTER 21

Into Space WE GO!

21.1 Introduction

Scientists in different fields have studied the nature of the universe for a very long time. But what about questions that we might not be able to answer, like what lies beyond the universe? The conversations about what lies beyond the universe can quickly become strange because we first have to define what "beyond the universe" means.

To define "beyond the universe" we need to define what we mean by the universe. Is the universe only what we can observe, or is it more than what we can observe? How can we know what is beyond what we can observe if we can't observe it? Will we be able to observe farther into the universe as technology advances?

Astronomers and physicists define the **observable universe** by the **speed of light**. Light includes all electromagnetic waves. Because we can only detect things that reflect or emit light, we cannot detect anything beyond the distance light can travel. As a result, the observable universe includes all the space and time light has traveled that we can detect.

And yet, evidence supports the idea that the universe is expanding. Is there a place for the universe to expand into? Is the place it expands into also the universe even though we can't see the place before the universe expands into it?

You can see how the questions get very strange quickly and how these questions bring up more questions than answers.

21.2 The Future of Space Research

Even if we can't know right now what lies beyond the observable universe, we live in an exciting time for space research. Science and technology have given us a variety of sophisticated tools like telescopes, space probes, rovers, and rockets to explore the observable universe. We have been able to put men on the moon, send rovers to Mars, land probes on comets, and send probes beyond the edge of our solar system. Many countries are involved in joint efforts to explore new ways to study space.

Lunar landing, 1969
Astronaut Charles Conrad Jr., Apollo 12 Commander

In the United States, cutting-edge projects in space exploration are pushing the boundaries of what's possible. NASA's **Artemis** program aims to return humans to the Moon, with a the goal of landing astronauts on the lunar surface in 2026. Artemis astronauts will include the first woman and the first man of color to land on the Moon.

This endeavor will utilize innovative technologies to explore more of the lunar surface than ever before and establish a sustainable presence as a stepping stone for future Mars missions.

The Dragonfly spacecraft
Artist's Concept

The Dragonfly mission is set to explore Saturn's moon Titan, using a drone-like **rotorcraft** to study prebiotic chemistry and habitability. Meanwhile, the **Psyche** mission will journey to a unique metal asteroid, offering insights into the building blocks of our solar system. On Mars, NASA's **Perseverance** rover is actively searching for signs of past life and collecting samples for future return to Earth. The rover has already made significant discoveries, including depositing sample tubes on the Martian surface that capture the planet's diverse geology. These sample tubes will be picked up and returned to Earth using unmanned spacecraft.

In addition to these missions, NASA's Space Launch System (SLS) is being developed as a heavy-lift launch vehicle capable of sending the Orion Multi-Purpose Crew Vehicle (MPCV) to a variety of deep-space destinations, including asteroids, the Moon, and ultimately Mars. The International Space Station (ISS) continues to be a hub of scientific discovery and international collaboration, now entering a decade focused on producing results from the orbiting laboratory. These projects represent just a fraction of the ongoing efforts in space exploration, with NASA and its partners working tirelessly to expand our understanding of the cosmos and our place within it.

In Europe, there are currently twenty-two countries exploring space by working together through the **European Space Agency (ESA)**. ESA also collaborates with NASA in various space missions, including the International Space Station (ISS), the SOHO satellite that makes solar observations, and the Hubble and James Webb Space Telescopes. ESA is also a contributor to NASA's Artemis mission, providing habitation and refueling modules and communications systems.

Launched in 2023, ESA's **Jupiter's Icy Moons Explorer (JUICE)** spacecraft will arrive at Jupiter in 2031, sending back observations of the giant gas planet along with its three large ocean-bearing moons, **Ganymede, Callisto,** and **Europa**. It will observe these moons as planetary objects and possible habitats.

Rosalind Franklin rover
Artist's Concept

In development is ESA's first Mars rover, **Rosalind Franklin**, that will drill 2 meters into the martian surface. Samples will be analyzed with an onboard lab, looking for organic material to reveal information about the early history of the planet.

ESA also has numerous other existing and planned missions, including the exploration of Mercury, mapping Earth's magnetic environment, monitoring Earth's sea-ice and vegetation, and looking for exoplanets.

This is an exciting time in astronomy as other countries join NASA and ESA in their search for new discoveries in our solar system and beyond.

21.3 The Future of Commercial Space Travel

Get ready for an adventure beyond Earth's atmosphere! The dream of traveling to Mars and beyond is becoming more tangible as advancements in space technology continue to progress. Historically, space exploration has been the domain of government agencies with the financial resources to undertake such ambitious projects. However, the landscape is changing as private companies increasingly enter the field, working on ways to provide commercial space travel to the public

A commercial spacecraft capable of carrying passengers approaches the ISS

The 21st century has seen the rise of the commercial space travel industry, with private companies developing technologies and services to make space more accessible. These companies are preparing for commercial space travel by developing spacecraft capable of carrying passengers into space, with some focusing on suborbital flights and others on orbital journeys. Suborbital flights are trips into space that do not require the spaceship to travel in orbit around Earth, while orbital flights involve sending a spacecraft into an orbit around Earth.

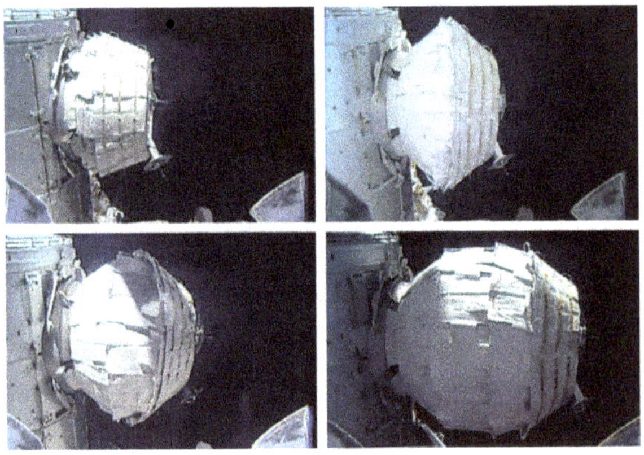

An experimental module being inflated at the ISS

Private companies are also partnering with government agencies to provide transport to the International Space Station and launching science experiments into space. They are also developing expandable modules for use as space stations and habitats on the Moon and Mars. These modules could potentially be rented for research purposes.

In addition to developing spacecraft and arranging flights, companies in the commercial space flight industry are producing needed fuels, tools, equipment, and supplies. Traveling into space requires the work of many people as well as advanced technology.

The commercial space industry is revolutionizing the global economy in ways that may be less obvious. Private companies have driven down launch costs to record lows, making it cheaper and easier to send thousands of satellites into orbit for a wide array of purposes.

This industry is expected to grow significantly in the coming years. With the right amount of money, anyone can buy a ticket to space. For half a million dollars, prosperous passengers can enjoy a suborbital trip. Or, for $50 million, upcoming orbital expeditions are on offer. Commercial spaceflight, or space tourism, is currently valued at more than $450 billion. Furthermore, it is expected to grow to more than $1 trillion by 2040.

In conclusion, the commercial space industry is making significant strides in making space travel more accessible to the public. While there are still many challenges to overcome, the progress made so far is promising and suggests that the dream of traveling to Mars and beyond may one day become a reality.

21.4 Summary

- Scientists are not sure what, if anything, lies beyond the **observable universe**.
- New technology is bringing us more discoveries in space.
- Commercial space travel is becoming a reality in the 21st century.

21.5 Vocabulary

Basic Words

Psyche (SIY-kee)

Titan (TIY-tuhn)

Challenge Words

Artemis (AHR-tuh-muhs)

Callisto (kuh-LIH-stoh)

Dragonfly (DRAA-guhn-fliy)

Europa (yoo-ROH-puh)

Ganymede (GAA-nih-meed)

Orion (uh-RIY-uhn)

Rosalind Franklin (RAH-zuh-lihnd FRAANK-luhn)

rotorcraft (ROH-tuhr-kraaft)

CHAPTER 22

Putting It All TOGETHER

22.1 Introduction

Science has opened the doors to incredible ways of seeing and understanding the world around us. Because of science we can probe the tiniest atomic objects and explore farther and farther into outer space. With science we can imagine new ways of travel, explore strange creatures in the deepest parts of the ocean, and begin to understand the nature of the electron. Most importantly, with science we can begin to solve real world problems, like cleaning up our environment, providing clean water for millions, and curing rare childhood illnesses.

But none of these things can happen without men and women working to discover the mysteries of the universe. Discoveries are what make new things possible, and yet discoveries start with basic questions like, "What is this made of?" "How far can this go?" "How does this work?" or "What happens if I change this?" Basic questions like these are the beginning of real discovery.

22.2 Amazing Discoveries

It's easy in the modern world to take for granted some of the amazing discoveries of the past. We forget that at one time what we now know about the world around us was not known.

In the chemistry section of this book you learned about how electrons are part of atoms and how atoms use electrons to form molecules. The idea of electrons gluing atoms together to make molecules and the scientific evidence to support this idea came from a series of small and very important discoveries.

In chemistry you also learned about heat and thermodynamics. Again, it was a series of small and very important discoveries that provided evidence to support the idea that there are laws that govern how heat is generated, converted, or lost, and that heat is a form of energy.

In biology, the task of organizing and naming all living things is a big job. Not only are there millions of different kinds of living things, but they can change over time, making the problem of sorting, naming, and understanding them that much harder. Some animals look very similar to each other and some look very different. Even those that may look similar don't necessarily belong in the same group. For example, roundworms and segmented worms look similar but are in different groups. Organizing living things and understanding their features is possible only because of discoveries and careful observation.

Discoveries in physics have made the impossible possible. Because we know how waves and electromagnetic radiation work, we can be sitting in one country and hear a concert happening in another country simply by turning on the radio or TV or using the internet. By knowing about waves, we can use a prism to split white light into many colors, build powerful telescopes, make microscopes with lenses, or have fun using mirrors to make us look fat, skinny, short, or tall.

Discoveries in physics also contribute to many discoveries in geology and astronomy. By using physics we can convert natural resources into usable energy for automobiles and stoves, and with physics we can explore deep space.

In geology, discoveries about the atmospheric cycle help us plan for changes in the weather and can help people survive extreme weather conditions. Knowing about how the different parts of ecosystems work together helps us understand how to reduce pollution and keep plants and animals healthy. In fact, knowing about all the different systems on Earth helps us understand that all the parts of Earth work together, and changing one thing may lead to unintended consequences.

With discoveries in geology we are able to find and extract fossil fuels for energy to heat our buildings and power our technology, and we know where to place solar panels, wind turbines, and dams to obtain electricity from renewable sources.

Astronomy is a rapidly advancing science. People have always studied the stars, but now astronomers can use advanced technology to gather lots of information and are constantly learning more and more about the universe. As more satellites, probes, telescopes, and rovers are put in space,

scientists are able to explore farther and farther into deep space. Discoveries in astronomy involve all the other fields of science. This requires **astrophysicists** to understand how the celestial bodies move and what path a spaceship needs to take. **Astrochemists** are analyzing what stars are made of. **Astrogeologists** are studying the structure of planets, comets, and asteroids. **Astrobiologists** are looking for life on other planets and ways to keep people alive in space and in habitats on other planets.

Discoveries, both small and large, have helped move us toward a technologically advanced world complete with cell phones, the internet, and possibly one day intergalactic space travel.

22.3 Famous Scientists

Figuring out how the world works, what it is made of, and what we can create is a lifelong passion for many scientists and engineers. Over the years many men and women have contributed to our understanding of the world around us through scientific investigation.

While all scientists contribute to our overall understanding of science, some scientists make such an important impact on what we know that they become famous!

One of the first early scientists to revolutionize the way we understand the world was **Nicolaus Copernicus** (1473–1543 C.E.). Copernicus is known as the father of modern astronomy and he was able to show conclusively that the Earth orbits the Sun. The idea that the Earth was not the center of the universe had been rejected for nearly 2000 years. **Aristarchus of Samos** (310–230 B.C.E.) was the first to consider the idea that the Earth orbited around the Sun and that the universe is expansive. However, no one believed him, and it wasn't until Copernicus published his work on a **heliocentric** cosmos that our understanding of the Earth, Sun, stars, planets, and greater universe changed.

Copernicus
1473–1543 C.E.

Aristarchus
310–230 B.C.E.

Albert Einstein
(1879–1955 C.E.)

Albert Einstein (1879–1955 C.E.) is one of the most famous scientists to date. His name is recognized by many people even if they have never studied science. Einstein changed our understanding of the laws of nature. In his day much of our understanding of how light, gravity, and time behaved came from Newton's laws. Einstein's most famous equation $E=mc^2$ showed how energy and matter can be converted into one another. This was revolutionary.

Three scientists, **James Watson**, **Francis Crick**, and **Rosalind Franklin** gave us the structure of DNA. Using X-ray **crystallography**, Rosalind Franklin (1920–1958 C.E.) mapped an image of DNA which aided James Watson (1928– C.E.) and Francis Crick (1916–2004 C.E.) in developing a model. The DNA molecule has revolutionized our understanding of living things and has led to amazing medical diagnoses and cures.

Rosalind Franklin
(1920–1958 C.E.)

22.4 The Nobel Prize

The **Nobel Prize** is the crown jewel of science. Most scientists, both young and old, dream of one day winning this award. The Nobel Prize is awarded yearly and given to the top scientists in physics, chemistry, and medicine, and additionally in the non-scientific fields of literature, peace, and economics.

The Nobel Prize medal

The Nobel prize was set up by **Alfred Nobel**, a Swedish chemist and inventor. Alfred Nobel invented dynamite but did not want to be remembered for this invention. Because he was very wealthy, he wrote in his will that the bulk of his estate was to be awarded to individuals who had contributed the "greatest benefit to mankind." Five years after Alfred Nobel's death, the Nobel Foundation was finally established and the first prizes awarded.

The winners of the Nobel Prize are determined by a committee of scholars who try to determine the most important work for the year. It is very difficult to win a Nobel prize. There have been some years when the Nobel Prize has not been awarded in a particular field because the committee did not find any discoveries they thought were important enough for the award.

Marie Curie
(1867–1934 C.E.)

Some famous scientists who have won the Nobel prize include Albert Einstein, **Niels Bohr**, **Richard Feynman**, and **Marie Curie**. In fact, Marie Curie was the first person in the history of the Nobel prize to be awarded two Nobel Prizes in science — one in chemistry and one in physics. In 1903 Marie Curie shared the Nobel Prize in physics with **Henri Becquerel** and **Pierre Curie** for the discovery of the elements **polonium** and **radium.** Then in 1911 Marie Curie won a second Nobel prize for her additional work that led to an understanding of the chemistry of polonium and radium.

The Nobel Prize has stirred the imaginations of many young scientists, and yet all Nobel Prize winners will probably say that spotting problems and asking good questions is the key to scientific discovery. Discoveries in chemistry, biology, physics, geology, and astronomy lead to new problems to be solved and new questions to ask. From these questions science, technology, and engineering move forward.

22.5 Summary

- Discoveries in science start with basic questions.
- Many amazing discoveries have revolutionized the way we think about the world.
- Many men and women have contributed to our understanding of science.
- The **Nobel Prize** is given to those whose work makes a significant contribution to their field of study.

22.6 Vocabulary

Basic Words

Becquerel, Henri (beh-KREHL, HEHN-ree)

Bohr, Niels (BAWR, NEELS)

Curie, Marie (KYOOR-ee, muh-REE)

Curie, Pierre (KYOOR-ee, pee-EHR)

Feynman, Richard (FIYN-muhn, RIH-chuhrd)

Nobel Prize (noh-BEHL, PRIYZ)

Challenge Words

Aristarchus of Samos (ehr-uh-STAHR-kuhs uhv SAY-mahs)

astrobiologist (aa-stroh-biy-AH-luh-jihst)

astrochemist (aa-stroh-KEH-mihst)

astrogeologist (aa-stroh-jee-AH-luh-jihst)

astrophysicist (aa-stroh-FIH-zuh-sihst)

Copernicus, Nicolaus (koh-PUHR-nuh-kuhs, NIH-koh-lahs)

crystallography (krih-stuh-LAH-gruh-fee)

Einstein, Albert (IYN-stiyn, AAL-fruhd)

heliocentric (hee-lee-oh-SEHN-trik)

polonium (puh-LOH-nee-uhm)

radium (RAY-dee-uhm)

Glossary / Index

(Pronunciation Key at end)

abdomen (AAB-duh-muhn) • a segment of an arthropod body, 56, 62, 64

absorb (uhb-SAWRB) • to take in, 18, 19, 31-33, 97-98, 101-103, 105, 110-111, 125, 153

acid rain (AA-suhd RAYN) • a type of air pollution, 151

acoustics (uh-KOO-stihks) • the study of sound, 98

air pressure • see atmospheric pressure

albedo (aal-BEE-doh) • the amount of light reflected by a surface, 125

alga (AAL-guh) [*plural,* **algae** (AAL-jee)] • a type of protist; usually found in water, 130-131, 135

algal bloom (AAL-guhl BLOOM) • an overgrowth of algae in a water ecosystem, 130-131, 135

alligator (AA-luh-gay-tuhr) • a type of reptile with scales and 4 legs, 74

aluminum (uh-LOO-muh-nuhm) • a metallic element, 96-97, 161

amphibian (aam-FIH-bee-uhn) • a type of vertebrate that usually begins life in water and as an adult lives on land and in water, 69, 73, 77

amplitude (AAM-pluh-tood) • the height of a wave measured from the middle, 81-84, 94

Animalia (aan-uh-MAH-lyuh) • the taxonomic category for animals, 40, 43

Annelida (aa-NEH-luh-duh) • the taxonomic category for segmented worms, 39-40, 48

antenna (aan-TEH-nuh) [*plural,* **antennae** (aan-TEH-nee)] • a sensory organ on the head of some arthropods, 57, 62

apex predator (AY-pehks PREH-duh-tuhr) • an organism that is at the top of the food chain and has no predator, 132-133

Apollo 11 mission (uh-PAH-loh ih-LEH-vuhn MIH-shuhn) • the NASA mission where the first people landed on the Moon, 159, 161

arachnid (uh-RAAK-nuhd) • a member of a group of arthropods that have 4 pairs of legs, 62

archaea (aar-KEE-uh) • a type of single-celled organism, 38, 40, 178

Aristarchus of Samos (aa-ruh-STAHR-kuhs uhv SAY-muhs) • Greek astronomer; found that Earth orbits the Sun and calculated the approximate distance between the Earth and the Moon, 169, 196

array (uh-RAY) • a unit made of a group of elements, 172, 183

Artemis (AHR-tuh-muhs) • a NASA mission to put humans on the Moon, 187, 188

by-product (BIY-prah-duhkt) • in a chemical reaction, a molecule or set of molecules that are produced in addition to the primary products, 150-152

Callisto (kuh-LIH-stoh) • a moon of Jupiter, 188

carbon (KAHR-buhn) • a type of atom, 11, 14, 117, 139-141, 144, 176-177

carbon-14 (KAHR-buhn fawr-TEEN) • a certain type of carbon atom, 144

carbon dioxide (KAHR-buhn diy-AHK-siyd) •a gas that is part of the carbon-oxygen cycle; a greenhouse gas; 117, 150, 153, 154

carbon-oxygen cycle (KAHR-buhn-AHK-sih-juhn SIY-kuhl) • the part of the atmospheric cycle in which animals inhale oxygen and exhale carbon dioxide and plants use the carbon from carbon dioxide to make food, and release oxygen into the air, 117

carnivore (KAHR-nuh vawr) • an animal that eats only other animals, 37, 132, 134

cartilage (KAHR-tuh-lihj) • a strong but lightweight and flexible tissue, 67

caterpillar (KAA-tuhr-pih-luhr) • a stage in the butterfly life cycle, 59-61

cephalothorax (seh-fuh-luh-THAWR-aaks) • a segment of some arthropod bodies that consists of the head and thorax fused together, 56, 62, 63, 64

chain reaction (CHAYN ree-AAK-shuhn) • a chemical or nuclear reaction resulting in energy or products that cause further reactions of the same kind, 145

chelicera (kih-LIH-suh-ruh) [*plural,* **chelicerae** (kih-LIH-suh-ree)] • a fang-like appendage on spiders, 62

chemical reaction (KEH-muh-kuhl ree-ACK-shuhn) • occurs when atoms join together, leave a molecule, or switch places, 2, 8-14, 22-34, 144-145, 150

chemistry (KEH-muh-stree) • the investigation of what physical things are made of and the ways in which they change; the study of matter, 2, 8-34, 149, 150, 193, 198

chitin (KIY-tuhn) • a type of polymer made with glucose molecules and nitrogen, 56, 62

chlorine (CLAWR-een) • a type of atom, 8-9, 13-14, 22-23

Chordata (kawr-DAH-tuh) • the taxonomic category for animals with a backbone, 39-40

chrysalis (KRIH-suh-luhs) • the hardened outer layer that protects a caterpillar undergoing metamorphosis, 60-61

cichlid (SIH-kluhd) • a type of small bony fish, 135

circular wave (SUHR-kyuh-luhr WAYV) • a wave where the molecules move outward from the center in a circular motion, 88, 90

climate (KLIY-muht) • the temperature, precipitation, wind, amount of sunshine, and other factors in a certain region averaged over a long period of time, 117-118, 134

closed system (KLOHZD SIH-stuhm) • one that is fully self-contained, 148

Cnidaria (niy-DAAR-ee-uh) • the taxonomic category that includes sea jellies and corals, 39, 40, 46

coal (KOHL) • a fossil fuel used for energy, 139-141, 150-151

compound eye (KAHM-pownd IY) • a type of eye made of many units; found in insects, 57

concentration (kahn-suhn-TRAY-shuhn) • in chemistry, the number of molecules in a particular space, 27-28

consumer, primary • see primary consumer

consumer, secondary • see secondary consumer

converging lens (kuhn-VUHRJ-ihng LEHNZ) • a lens that bends the incoming light so the light rays coming out of the opposite side of the lens converge, or meet, at a single point, 114-115

Copernicus, Nicolaus (koh-PUHR-nuh-kuhs, NIH-koh-lahs) • showed conclusively that Earth orbits the Sun, 196

covalent bond (koh-VAY-luhnt BAHND) • a type of chemical bond in which electrons are shared equally, 13-14

crab (CRAAB) • a type of arthropod in the group crustaceans, 63

Crick, Francis (KRIHK, FRAAN-suhs) • one of the developers of the DNA model, 197

crude oil (KROOD OYL) • oil as it come directly from the ground, 139

crust (KRUHST) • in geology, the hard, rocky outer layer of Earth, 117, 123

crustacean (kruh-STAY-shuhn) • a group of arthropods that includes animals that have 5 pairs of legs, 63

crystal (KRIH-stuhl) • a solid that has a highly organized lattice structure, 18

crystallography (krih-stuh-LAH-gruh-fee) • the branch of science that studies the structure of crystals, 197

Curie, Marie (KYOOR-ee, muh-REE) • won 2 Nobel Prizes in science, 198

Curie, Pierre (KYOOR-ee, pee-EHR) • winner of a Nobel Prize, 198

Curiosity (kyoor-ee-AH-suh-tee) • a Mars rover, 179

cycle (SIY-kuhl) • a series of events that repeat, 4, 84, 117-126

decibel (DEH-suh-behl) • a unit of measure of sound, 95

decomposer (dee-kuhm-POH-zuhr) • an organism that consumes dead plants and animals, 133

deforestation (dee-fawr-uh-STAY-shuhn) • the cutting or burning down of forests, 154

deplete (dih-PLEET) • to use up, 141

diffract (dih-FRAAKT) • in physics, for waves or light to change direction, travel around corners, travel around obstacles, or move through openings, 97, 101

discovery (dihs-KUH-vuh-ree) • in science, understanding or seeing something previously not understood about how the world works, 1-7

displace (dihs-PLAYSS) • to move out of position, 85-87

distill (dih-STIHL) • purify, 139

distillation (dih-stuh-LAY-shuhn) • a process for purifying liquids, 139

diverging lens (duh-VUHRJ-ihng LEHNZ) • a lens that bends the incoming light so the light rays coming out of the opposite side of the lens diverge, spreading the light rays away from each other, 114, 115

domain (doh-MAYN) • in biology, a taxonomic category, 38, 40

echinoderm (ih-KIY-nuh-derm) • an organism in the phylum Echinodermata, 51-53

Echinodermata (ih-kiy-nuh-DEHR-muh-tuh) • the taxonomic category that includes sea stars and sea urchins, 39-40, 51-53

echo (EH-koh) • the repetition of a sound when sound waves are reflected, or bounced back from a surface, 98

ecological community (ee-kuh-LAH-jih-kuhl kuh-MYOON-uh-tee) • all the living things in an ecosystem, 129-130

ecological niche (ee-kuh-LAH-jih-kuhl NIHCH) • the particular place within an ecosystem that an organism occupies, 133-136

ecology (ih-KAH-luh-jee) • the study of the relationships between living things and their environment, 128-136

ecosystem (EE-koh-sihs-tuhm) • a community of living things and their surroundings that function as a unit, 5, 129-131, 133, 135-136, 148, 155

Einstein, Albert (IYN-styn, AAL-buhrt) • physicist who showed how energy and matter can be converted into one another, 196

electromagnetic radiation (ih-lehk-troh-maag-NEH-tihk ray-dee-AY-shuhn) • energy that is given out, or radiated, from its source by electromagnetic waves, 103-104, 107

electromagnetic spectrum (ih-lehk-troh-maag-NEH-tihk SPEHK-truhm) • the full range of electromagnetic waves, 89, 104-105

electromagnetic wave (ih-lehk-troh-maag-NEH-tihk WAYV) • a combination of electric and magnetic fields that can transport energy through a vacuum, 89, 90, 103-104, 172, 174, 186

electron (ih-LEHK-trahn) • one of the 3 basic parts of an atom; found on the outside of the atom, 8-14, 22, 193

El Niño (EHL NEE-nyoh) • a cyclical weather pattern, 120

emit (EE-miht) • to give or send out, 106, 110, 186

endothermic (ehn-doh-THUHR-mihk) • a chemical reaction that absorbs heat energy, 31-32

energy (EH-nuhr-jee) • something that gives something else the ability to do work, 17-20, 25, 30-34, 85-90, 95-96, 102-103, 105-106, 124-126, 131, 133, 138-145, 148, 153, 193-194

energy cycle (EH-nuhr-jee SIY-kuhl) • the geological cycle in which energy is reflected and absorbed by Earth's surface and drives all the other cycles, 117, 124-126, 131, 148

environment (ihn-VIY-ruhn-muhnt) • everything that surrounds a living thing in the area where it lives, 128-129, 142

ESA (EE-SUH) • the European Space Agency, 171, 188-189

Eukarya (yoo-KAA-ree-uh) • the taxonomic domain for animals, 38, 40

Europa (yoo-ROH-puh) a moon of Jupiter 178, 188

evaporate (ih-VAA-puh-rayt) • to turn a liquid into a gas, 19

exoplanet ((EHK-soh-plaa-nuht) • a planet that exists outside our solar system, 180-184

exoskeleton (ehk-soh-SKEH-luh-tuhn) • the tough outer shell of an arthropod, 56, 57, 62, 63

exothermic (ehk-soh-THUHR-mihk) • a chemical reaction that releases heat energy, 31

extremophile (ihk-STREE-muh-fiy-uhl) • an organism that lives under extreme conditions, 178

Feynman, Richard (FIYN-muhn, RIH-chuhrd) • physicist who won a Nobel Prize, 198

First Law of Thermodynamics • states that energy cannot be created or destroyed, 32, 34

flagellum (fluh-JEH-luhm) [*plural,* **flagella** (fluh-JEH-luh)] • a whiplike appendage on some cells, 45

flatworm (FLAAT-wuhrm) • a long, thin worm in in the phylum Platyhelminthes, 39, 47

fluorine (FLAWR-een) • a type of atom, 12, 13

food chain (FOOD CHAYN) • the order in which the organisms in an ecosystem use each other as a food source, 131-133

force (FAWRSS) • any action that changes the location of an object, the shape of an object, how fast or how slowly an object is moving, 17, 102, 119, 121

fossil (FAH-suhl) • an organism that has been converted to minerals over a very long time, 139-141

fossil fuel (FAH-suhl FYOOL) • gasoline, coal, natural gas, 139-141, 151-152, 154

fraction (FRAAK-shuhn) • a group of hydrocarbons that have similar chemical properties, 140

frame (FRAYM) • the main body of a rocket, 161

Franklin, Rosalind (FRAANK-luhn, RAH-zuh-lihnd) • mapped an image of DNA, 197

frequency (FREE-kwuhn-see) • a measure of how fast a wave moves, 81-84, 93-95, 103-104

frog (FRAHG) • a type of amphibian that undergoes metamorphosis, 69-72, 77

fuselage (FYOO-suh-lahzh) • the main body of an aircraft; accommodates the crew and passengers or cargo, 161

GAIA (GIY-uh) • a star mapping satellite, 171

gamma ray (GAA-muh ray) • a type of electromagnetic wave, 89, 104, 172

Ganymede (GAA-nih-meed • a moon of Jupiter, 188

gas (GAAS) •

1) a state of matter in which the substance does not have a definite shape and can expand indefinitely, 16-22, 153-154
2) a fossil fuel; gasoline, 138-140, 150-151

generator (JEH-nuh-ray-tuhr) • a machine that produces electricity, 141-142, 146

geological cycle (gee uh-LAH-juh-kuhl SIY-kuhl) • one of several cycles that affect Earth's geology, 117

geology (jee-AH-luh-jee) • the study of Earth, 2, 4-5, 117- 156, 194

glucose (GLOO-kohs) • a type of sugar molecule, 56

greenhouse effect (GREEN-howss ih-FEHKT) • the warming of Earth due to the trapping of the Sun's energy by certain gases, 153-154

greenhouse gas (GREEN-howss GAAAS) • a gas in the atmosphere that traps heat from the Sun, 153-154

Grinnell, Joseph (gruh-NEHL JOH-zuhf) • the naturalist who was first to use and define the word niche, 133

guidance system (GIY-duhns SIHS-tuhm) • the equipment that guides a rocket, 162

habitable zone (HAA-buh-tuh-buhl ZOHN) • an area that is the right distance from its sun that a planet might be able to support life as we know it, 180-182

heat (HEET) • a form of energy that happens when atoms and molecules move, 18-19, 25-26, 29-34, 105, 110, 119-122, 125, 153-154, 193

heat budget (HEET BUH-juht) • the balance between Earth's incoming and outgoing heat, 125

helium (HEE-lee-uhm) • a type of atom, 10

herbivore (UHR-buh-vawr) • an animal that eats only plants, 37, 131-132, 134

Hutton, William (HUH-tuhn, WIHL-yuhm) • English geologist; found that coal is made of plant fossils, 141

hydrocarbon (HIY-droh-kahr-buhn) • a material made of hydrogen and carbon, 139-140

hydroelectric (hiy-droh-ih-LEHK-trihk) • relating to the production of electricity by waterpower, 141-142

igneous rock (IHG-nee-uhs RAHK) • formed when molten magma inside Earth cools and hardens, 121, 124

infinite (IHN-fuh-nuht) • having no limit or end, 173

infrared (ihn-fruh-REHD) • a type of electromagnetic wave, 89, 104, 172

insect (IHN-sehkt) • a type of arthropod, 39, 56-62, 64

instrument unit (IHN-struh-muhnt YOO-nuht) • in a rocket, contains the devices needed to operate it, 162

intensity (ihn-TEHN-suh-tee) • in sound, loudness or softness, 94-95

interdependent (ihn-tuhr-dih-PEHN-duhnt) • depending on one another, 134

invertebrate (ihn-VUHR-tuh-bruht) • an animal that does not have a backbone, 43-64

ionic bond (iy-AH-nihk BAHND) • a type of chemical bond in which electrons are not shared, 13-14, 22, 23

iron oxide (IY-uhrn AHK-siyd) • a type of molecule, 24

ISS (IY-EHS-EHS) • the International Space Station; a manned satellite that orbits Earth, 163, 187, 188

jellyfish • see sea jelly

kerosene (KEHR-uh-seen) • a hydrocarbon fuel, 140, 162

Lake Victoria (LAYK vihk-TAWR-ee-uh) • the second largest freshwater lake, 135

lamprey (LAAM-pree) • a type of jawless fish, 67

lattice (LAA-tuhs) • in physics, the organization of molecules in a particular repeating pattern, 17-18

law of reflection (LAW uhv rih-FLEHK-shuhn • states that the angle of reflected light will be equal to the angle of the incoming light, 112-113, 115

law of unintended consequences (LAW uhv uhn-ihn-TEHN-duhd KAHN-suh-kwehn-suhz) • unexpected and undesirable results when part of a complex system is changed, 135-136

Lepidoptera (leh-puh-DAHP-tuh-ruh) • the group of insects that has scaled wings and includes butterflies, 58

light (LIYT) • electromagnetic radiation, 84, 89, 101-115, 125-126, 142-143, 186

liquid (LIH-kwuhd) • a state of matter; a substance that flows and takes the shape of its container, 16-20, 33, 162-165, 177

liquid hydrogen (LIH-kwuhd HIY-druh-juhn) • a type of rocket fuel, 162

liquid oxygen (LIH-kwuhd AHK-suh-juhn) • a type of rocket fuel, 162

lithium (LIH-thee-uhm) • a type of atom, 10, 13, 106

lizard (LIH-zuhrd) • a type of reptile with scales and 4 legs, 74

lobster (LAHB-stuhr) • a type of arthropod in the group crustaceans, 39, 56, 63

longitudinal wave (lahn-juh-TOOD-nuhl WAYV) • a wave where the molecules are displaced parallel to the direction the wave is moving, 87-88, 90, 92, 95

mammal (MAA-muhl) • a type of vertebrate that has hair and mammary glands, 75-76

mammary gland (MAA-muh-ree GLAAND) • in mammals, a gland that produces milk, 76

mandible (MAAN-duh-buhl) • a mouthpart of an arthropod, 57

mechanical wave (mih-KAA-nih-kuhl WAYV) • a wave that needs some type of matter to move through, 88-89, 90, 92, 95, 99

medium (MEE-dee-uhm) • in physics, the material a sound wave travels through, 92-94, 114

metamorphic rock (meh-tuh-MAWR-fihk RAHK) • rock that has changed from one type of rock to another as a result of very high heat and pressure, 121, 123-124

metamorphosis (meh-tuh-MAWR-fuh-suhs) • the process of transforming from one form to a completely different form, 61, 71-72

methane (MEH-thayn) • a type of molecule; greenhouse gas, 14, 140, 153

molecule (MAH-lih-kyool) • two or more atoms hooked together, 2, 8-34, 85-88, 90, 92, 94-96, 111, 119-120, 150, 193

Mollusca (muh-LUHS-kuh) • the taxonomic category that includes snails and octopuses, 39-40, 48-50

mollusk (MAH-luhsk) • an animal in the phylum Mollusca, 48-50

molt (MOHLT) • to shed a too small outer skin so a new skin can grow, 60

motor (MOH-tuhr) • a machine that converts electrical energy into mechanical energy, 142

NASA (NAA-suh) • the U.S. National Aeronautics and Space Administration; develops and oversees space missions, 159, 161, 163, 179, 187-188

natural gas (NAA-chuh-ruhl GAAS) • a hyrdocarbon used for fuel and heat, 139-140

natural resources (NAA-chuh-ruhl REE-sawr suhz) • materials used by humans and found in or on Earth in a natural state, 138-146

Nematoda (nehm-uh-TOH-duh) • the taxonomic category for roundworms, 39-40, 48

neon (NEE-ahn) • a type of atom, 12

neutron (NOO-trahn) • one of the 3 basic parts of an atom; found in the center of the atom, 9, 14, 144-145

Nile perch (NIYL PUHRCH) • a species of fish, 135

nitrogen (NIY-truh-jehn) • a type of atom, 11, 56, 117, 144, 150-151

Nobel Prize (noh-BEHL PRIYZ) • awarded to top scientists in several disciplines, 197-198

nuclear (NOO-klee-uhr) • in physics, relating to the nucleus of an atom, 144-145

nuclear energy (NOO-klee-uhr EH-nuhr-jee) • produced by a nuclear reaction, 144-145

nuclear fission (NOO-klee-uhr FIH-shuhn) • the splitting of an atom into other elements during a nuclear reaction, 144

nuclear reaction (NOO-klee-uhr ree-AAK-shuhn) • a reaction in which the protons and neutrons get moved in and out of an atomic nucleus, changing the element, 144-145

nuclear reactor (NOO-klee-uhr ree-AAK-tuhr) • a large device that produces nuclear energy, 144-145

nucleus (NOO-klee-uhs) •

1) in biology, the part of an animal cell that contains the DNA, 38
2) in physics, the part of an atom that contains protons and neutrons, 144

observable (uhb-ZUHR-vuh-buhl) • able to be seen or detected, 186

octopus (AHK-tuh-puus) • an invertebrate animal in the phylum Mollusca, 48, 50

omnivore (AHM-nih-vawr) • an animal that eats both plants and animals, 37

oral arm (AW-ruhl AHRM) • appendage on a sea jelly used to sweep food into the mouth, 46

orbiter (AWR-buh-tuhr) • a spacecraft designed to orbit a body in space, 163

organ (AWR-guhn) • in biology, a structure that is made of tissues and that performs a specific function, 36-37

oscillate (AH-suh-layt) • to move back and forth, 83-84, 88, 93

oxygen (AHK-sih-juhn) • a type of atom, 11, 24, 37, 117, 135, 150-152, 162-163, 177-178

ozone (OH-zohn) • a molecule made of 3 oxygen atoms, 152

parallax (PEHR-uh-laaks) • the effect that makes a nearby object appear to shift position when it is viewed from different locations, 169-171

parallel (PEHR-uh-lehl) • everywhere equally distant, 87, 90, 115

parasite (PEHR-uh-siyt) • an organism that lives inside another organism and feeds on it, 47-48

particle (PAHR-tih-kuhl) • a tiny bit of something; one of the basic units of matter, 89, 93, 101-103, 109, 111-112

payload (PAY-lohd) • the load carried on a rocket, 161

peak (PEEK) • in physics, the height of a wave, 80-84, 90, 93

pedipalp (PEH-duh-paalp) • an appendage on spiders that is used to grab and position prey, 62

perpendicular (puhr-puhn-DIH-kyuh-luhr) • at a right angle, 86, 87

Perseverance (puhr-suh-VIHR-ehns) • a Mars rover, 179-180

photon (FOH-tahn) • a packet of energy that has a particular frequency and travels through space like a wave, 103-104, 106, 107

photosynthesis (foh-toh-SIHN-thuh-suhs) • the process by which plants convert solar energy in the form of light into chemical energy in the form of food, 125-126

photovoltaic cell (foh-toh-vahl-TAY-ihk SEHL) • a device that releases electrons when light energy strikes it, 142-143

phylum (FIY-luhm) [*plural,* **phyla** (FIY-luh)] • a taxonomic category, 38-40, 44, 46, 47, 48, 51

physics (FIH-zihks) • the branch of science that explores how the natural world works, 2, 3-4, 79-116, 194, 198

pitch (PIHCH) • how high or low a sound is, 84, 93, 99

plankton (PLAANK-tuhn) • tiny organisms that live in water, 46

Platyhelminthes (plaa-tih-hehl-MIHN-theez) • the taxonomic category for flatworms, 39-40, 47

pollution (puh-LOO-shuhn) • the introduction of man-made waste that makes an environment unsuitable or unsafe for use, 150-152

polonium (puh-LOH-nee uhm) • an element, 198

Porifera (puh-RIH-fuh-ruh) • the taxonomic category that contains sponges, 39-40, 44-45

precipitation (prih-sih-puh-TAY-shuhn) • rain, sleet, hail, snow, 118, 120

predator (PREH-duh-tuhr) • an organism that primarily obtains food by killing and consuming other organisms, 132-133, 134

pressure (PREH-shuhr) • the measure of the number of molecules and how fast they are moving in a given space, 119-120, 121, 122, 123

primary consumer (PRIY-meh-ree kuhn-SOO-muhr) • an herbivore, 132

primary product (PRIY-meh-ree PRAH-duhkt) • in a chemical reaction, the main molecule or set of molecules that are produced, 150

producer (pruh-DOO-suhr) • organisms in the first trophic level that make food from the Sun's energy, 131

propulsion system (pruh-PUHL-shuhn SIHS-tuhm) • in a rocket, provides the power, 162

proton (PROH-tahn) • one of the 3 basic parts of an atom; found in the center of the atom, 9, 14, 144

Psyche (SIY-kee) • a space mission, 187

pumice (PUH-muhs) • a type of volcanic rock that is light and full of air holes, 122

quantum mechanics (KWAHN-tuhm mih-KAA-nihks) • includes the theory that photons are the fundamental unit of light, 103

radiate (RAY-dee-ayt) • to send out rays, 103-104

radioactive (ray-dee-oh-AAK-tihv) • an element that emits energetic particles by disintegrating its atomic nucleus, 103, 144

radio telescope (RAY-dee-oh TEH-luh skohp) • an instrument that uses radio waves to study objects in space, 172, 183

radio wave (RAY-dee-oh WAYV) • a type of electromagnetic wave, 89, 104, 172, 183

radium (RAY-dee-uhm) • an element, 198

radula (RAAD-juh-luh) • a spiky tongue in some mollusks, 49

recycle (ree-SIY-kuhl) • to process materials in order to regain substances for human use, 149-150

refine (rih-FIYN) • purify, 139

reflect (rih-FLEHKT) • to throw or bounce back, 97-98, 101, 110-115, 125, 181

refract (rih-FRAAKT) • to cause a wave to bend or change speed, 97, 105, 110, 113-115

renewable fuel (rih-NOO-uh-buhl FYOOL) • one that is capable of being replaced by natural processes, 141-144

reptile (REHP-tiyl) • a member of the group of vertebrate animals that use the Sun's energy to regulate their body temperature, 73-74

rigid (RIH-juhd) • inflexible; not easy to deform or bend, 96

rock cycle (RAHK SIY-kuhl) • the geological cycle in which rocks are continuously created, transformed, broken down, and created again, 117, 121-124, 126

rocket (RAA-kuht) • a type of engine used to launch objects into space, 161-163

Rosalind Franklin (RAH-zuh-lihnd FRAANK-luhn) • ESA's Mars rover, 188

rotorcraft (ROH-tuhr-kraaft) • an aircraft that flies by use of rotating blades, 187

roundworm (ROWND-wuhrm) • a small, smooth, round worm in the phylum Nematoda, 39, 47-48

sand dollar (SAAND DAH-luhr) • an invertebrate organism in the phylum Echinodermata, 51, 52

Saturn V (SAA-tuhrn FIYV) • a rocket that was designed to carry people and heavy equipment, 161-163

scatter (SKAA-tuhr) • to go in many different directions at once, 111-112, 115

scavenger (SKAA-vuhn-juhr) • an animal that eats dead animals, 132

sea jelly (SEE JEH-lee) • an invertebrate in the phylum Cnidaria; a jellyfish, 46

sea star (SEE STAHR) • an invertebrate organism in the phylum Echinodermata; also called a starfish, 51, 52

sea urchin (SEE UHR-chuhn) • an invertebrate organism in the phylum Echinodermata, 51, 52-53

secondary consumer (SEH-kuhn-dehr-ee kuhn-SOO-muhr) • a carnivore that eats herbivores, 132

sedimentary rock (seh-duh-MEHN-tuh-ree RAHK) • formed from bits of rock and other materials left behind by wind and water and subjected to intense pressure and heat, 121-124, 141

segment (SEHG-mehnt) • a section of something, 48

segmented worm (SEHG-mehn-tuhd wuhrm) • a worm that has a segmented body; in the phylum Annelida, 47-48

shark (SHAHRK) • a type of cartilage fish, 67, 129

sheath (SHEETH) • in some insects, a covering for the wings, 57

tadpole (TAAD-pohl) • the early stage of a frog, 69-72

tentacle (TEHN-tih-kuhl) • a long, stringy appendage on some animals, 46

thermodynamics (thur-moh-DIY-naa mihks) • the study of heat, 29, 32, 34

thorax (THAWR-aaks) • a segment of an arthropod body, 56, 57, 62

threshold (THRESH-hohld) • the level or point above which something will take place and below which it will not, 95

tidal power (TIY-duhl POW-uhr) • use of the movement of water in ocean tides to run turbines, 142

tissue (TIH-shoo) • in a living organism, a group of cells that perform a particular function, 36-38

titanium (tiy-TAY-nee uhm) • a metallic element, 161

transfer (TRAANS-fuhr) • to pass from one to another, 32-33, 85, 87, 90, 96

transform (traans-FAWRM) • to change, 61, 121, 123, 124, 126

transit (TRAAN-suht) • happens when a planet crosses in front of its sun, 182

transmit (traans-MIHT) • in physics, to move through a medium, 97, 110, 112, 115

transparent (traans-PEHR-uhnt) • describes an object or material that can be seen through, 57

transverse wave (traans-VUHRS WAYV) • a wave in which the molecules in a material move perpendicular to the direction the wave is traveling, 86-87, 90

trophic level (TROH-fihk LEH-vuhl) • the position an organism occupies in a food chain, 131-133, 136

turbine (TUHR-buhn) • a machine that converts the flow of liquid or air into work, 141-145

turtle (TUHR-tuhl) • a type of reptile that has a shell covering the body, 74

umbrella (uhm-BREH-luh) • in biology, a sea jelly's body; also called the bell, 46

unsheathed (uh-SHEETHD) • insect wings that do not have a sheath, 57

valley (VAA-lee) • in physics, the lowest part of a wave, 80

vertebra (VUHR-tuh-bruh) [*plural,* **vertebrae** (VUHR-tuh-bray)] • one of the several bones that make up the backbone, 43

vertebrate (VUHR-tuh-bruht) • an animal that has a backbone, 43-44, 66-77

vibration (viy-BRAY-shuhn) • in physics, the back and forth movement of particles in a medium, 92-94

visible light (VIH-zuh-buhl LIYT) • that part of the electromagnetic spectrum that we can see with our eyes, 89, 104-105, 110-112

Pronunciation Key

aa	past	(PAAST)
ay	date	(DAYT)
ah	far	(FAHR)
aw	saw	(SAW)
eh	pet	(PEHT)
ee	feed	(FEED)
ih	trip	(TRIHP)
iy	side	(SIYD)
oh	bone	(BOHN)
oi	boy	(BOI)
oo	rule	(ROOL)
ow	noun	(NOWN)
uh	fun	(FUHN)
uu	wood	(WUUD)
y	yet	(YEHT)

Image Credits

Cover and Title Page: Photo by artegorov3@gmail, AdobeStock
Chapter 1: Page 3: By Ermolaev Alexandr, AdobeStock
Chapter 3: Page 16: By MF, AdobeStock
Chapter 4: Page 24: Top, Public Domain; Bottom, By slay19, AdobeStock
Chapter 5: Page 29: By ulga, AdobeStock; Page 30: Top, By masyastadnikova, AdobeStock; Bottom, By Africa Studio, AdobeStock; Page 31: By zmijak, AdobeStock
Chapter 6: Page 36: 1&2. Public Domain; 3. By Dr_Microbe, AdobeStock; 4&5. By Digital Photo, AdobeStock; Page 37: By zcy, AdobeStock; Page 38: 1. By David Ellifrit, NOAA NMFS/AKFSC; 2. By Charles J Sharp, CC BY SA 4.0; 3. By Christel SAGNIEZ from Pixabay; 4. By Andreas Eichler-CC BY SA 3.0; Animal Cell, By Andrea Danti, AdobeStock
Chapter 7: Page 43: By jim, AdobeStock; Page 45: 1-3. NOAA; 4. Bernard Picton, CC BY SA 4.0; 5. By Alcinoe at English Wikipedia, Public Domain; Page 46: By rost9, AdobeStock; Page 47: 1. By Richard Ling, CC BY SA 2.0; 2. By wernerrieger, AdobeStock; 3&4. Photo Collection of Dr. James P. McVey, NOAA Sea Grant Program; 5. By David Burdick, NOAA; 6. By Roziline, AdobeStock; Page 48: Top, By sinhyu, AdobeStock; Bottom, By Bohdan, AdobeStock; Page 49: 1. By Kaen S., AdobeStock; 2. By Sylke Rohrlach from Sydney, CC BY SA 2.0; 3. By Daniel Lamborn, AdobeStock; 4. By Nick Hobgood, AdobeStock; 5. By Chika Watanabe from Los Altos, USA, CC BY SA 2.0; 6. By cherylvb, AdobeStock; 7. By diveivanov, AdobeStock; 8. By Vladislav, AdobeStock; Page 50: 1. By NURC/UNCW and NOAA/FGBNMS; 2. By albert kok, CC BY SA 3.0; 3. By Daiju Azuma from KOBE, Japan, CC BY SA 2.5; 4. Photo by Vlad Tchompalov on Unsplash; Page 51: 1. Photographer, Julie Bedford, NOAA PA; 2. Andrew David, NOAA/NMFS/SEFSC Panama City; Lance Horn, UNCW/NURC-Phantom II ROV operator; 3. Photographer, Claire Fackler, CINMS, NOAA; 4. Olympic Coast National Marine Sanctuary, Public Domain; 5. By Smart Destinations, CC BY SA 2.0; 6. Public Domain; Page 52: 1. By Frédéric Ducarme, CC BY SA 4.0; 2. By Didier Descouens, CC BY SA 3.0; 3. By David Monniaux, CC BY SA 3.0; 4. By Jerry Kirkhart from Los Osos, Calif, CC BY SA 2.0; 5. By Didier Descouens-CC BY SA 4.0
Chapter 8: Page 58: Butterfly, By Didier Descouens, CC BY SA 4.0 International; Closeup, By H. Zell, CC BY SA 3.0; Page 59: 1. By christian vinces, AdobeStock; 2. By Roberta Canu, AdobeStock; 3. By michaklootwijk, AdobeStock; 4. By qaz1235, AdobeStock; Page 60: 1. By Ivar Leidus, CC BY SA 4.0 International; 2. By Atanu Bose Photography, CC BY SA 4.0 International; 3. By Sourabh. biswas003, CC BY SA 4.0 International; 4. By Hamilton, AdobeStock; 5. By Yvonne, AdobeStock; Page 61: 1. By David Sussman, AdobeStock; 2. By leekris, AdobeStock; 3. By Kenneth Dwain Harrelson, CC BY SA 3.0; Page 63: Lobster, By Lsantilli, AdobeStock; Shrimp, By showcake, AdobeStock; Crab, By xamtiw, AdobeStock
Chapter 9: Page 67: Top: NOAA Great Lakes Environmental Research Laboratory, CC BY SA 2.0; Bottom: by Amada44, CC BY SA 3.0; Page 68: Public Domain; Page 69: 1. By David Mark from Pixabay; 2. By Rushenb, CC BY SA 4.0; 3. By Alexas Fotos from Pixabay; Page 70: Top, By Marc Andreu, AdobeStock; Bottom: By +NatureStock, AdobeStock; Page 71: Top, By Ruckszio, AdobeStock; Bottom: By kuritafsheen, AdobeStock; Page 72: Top, By kuritafsheen, AdobeStock; 1. By kikkerdirk AdobeStock; 2. By Michael Gäbler, CC BY SA 3.0; 3. By LiquidGhoul edited by Muhammad, CC BY SA 3.0; 4. By Ivan Kuzmin, AdobeStock; 5. By Charles J. Sharp, CC BY SA 4.0 International; 6. By Holger Krisp, CC BY SA 3.0; Page 73. Top, By Rjcastillo, CC BY SA 3.0; Bottom, By JeanPierre, AdobeStock; Page 74: Lizard, By bennytrapp, AdobeStock; Crocodile, By fotoslaz, AdobeStock; Turtle, By Eric Isselée, AdobeStock; Page 75: 1. By Steve Byland, AdobeStock; 2. By Markus De Nitto from Pixabay; 3. By Zdeněk Macháček on Unsplash; 4. By Catherine Merlin on Unsplash; Page 76: 1. NASA; 2. By Charles J. Sharp, CC BY SA 4.0; 3. By Lukas, AdobeStock; 4. By Graeme Rainsbury, Public Domain; 5. By Nils Bouillard on Unsplash; 6. By Rasbak, CC BY SA 3.0; 7. By ANADMAN, AdobeStock
Chapter 11: Page 93: Left, By phaitoon, AdobeStock Right, By nataba, AdobeStock;
Chapter 12: Page 102: Public Domain; Page 106: AnnaLangova, Public Domain
Chapter 13: Page 111: By Aaron DeLonay, USGS; Page 113: By Ruslan Russland, AdobeStock
Chapter 14: Page 120: NASA; Page 122: Top, NPS, Public Domain; Bottom, By Tonya Troxler, USGS; Page 123: By Jan Canty on Unsplash
Chapter 15: Page 129: Image by 5406753 from Pixabay; Page 130: Top, By henk bogaard, AdobeStock Bottom, USGS/Dr. Jennifer L. Graham
Chapter 16: Page 142: By Ryan Thorpe on Unsplash; Page 143: Top, USAF, Nellis Air Force Base; Bottom, DOE/NREL, by Warren Gretz
Chapter 17: Page 148: NASA, Public Domain; Page 149: By Ulrich Müller, AdobeStock 151: By Tomasz Bidermann, AdobeStock; Page 152: By Marina Shemesh, Public Domain; Page 154: Top, NOAA; Bottom, By whitcomberd, AdobeStock
Chapter 18: Page 159-164: NASA, Public Domain
Chapter 19: Page 170: By Mykola Velychko, AdobeStock; Page 171: Artist's impression/ESA; Page 172: Top: NRAO/AUI and NRAO/AUI Photographer, Bob Tetroi; Bottom: NASA and ESA, Hubble Space Telescope; Page 173: By Christos Georghiou, AdobeStock
Chapter 20: Page 177: Moon, NASA; Mercury, JPL and NASA/Johns Hopkins University Applied Physics Laboratory/Carnegie Institution of Washington; Venus, NASA/JPL-Caltechz; Page 178: Europa, NASA/JPL/DLR; Europa detail, NASA/JPL/University of Arizona; Mars, NASA; Page 179: NASA/JPL-Caltech; Page 180, 181, 182: NASA; Page 183: By CGP Grey, CC BY SA 2.0
Chapter 21: Page 186: NASA; Page:187: Artist's Concept/NASA APL; Page 188: Artist's Concept, ESA Medialab; Page 189: NASA; Page 190: NASA
Chapter 22: Page 196: Public Domain; Page 197: Both-Public Domain; Page 198: Public Domain

www.ingramcontent.com/pod-product-compliance
Lightning Source LLC
Jackson TN
JSHW072152080525
83703JS00001B/7
9781941181881